Talking with Patients

MIT Press Series on the Humanistic and Social Dimensions of Medicine

Stanley Joel Reiser, General Editor

Talking with Patients

Volume 2:
Clinical Technique

Eric J. Cassell

The MIT Press
Cambridge, Massachusetts
London, England

This book was set in Baskerville by Asco Trade Typesetting Ltd., Hong Kong, and printed and bound by The Murray Printing Company in the United States of America

Library of Congress Cataloging in Publication Data

Cassell, Eric J., 1928–
  Talking with patients.

  (MIT Press series on the humanistic and social dimensions of medicine; 1–2)
  Includes bibliographies and indexes.
  Contents: v. 1. The theory of doctor-patient communication—v. 2. Clinical technique.
  1. Physician and patient. 2. Interpersonal communication. 3. Medical history taking. 4. Medicine, Clinical. I. Title. II. Series. [DNLM: 1. Communication. 2. Physician-Patient Relations. W1 MI938M v. 1–2/ W 62 C344t]
  R727.3.C38   1985   610.69′6   84-26120

Volume 1
ISBN 0-262-03111-6 (h)
      0-262-53055-4 (p)
Volume 2
ISBN 0-262-03112-4 (h)
      0-262-53056-2 (p)

To my beloved and loving children,
Stephen and Justine

Contents

Medical encounters begin with dialogue. Language transforms the experience of illness into subjective portraits painted by the patient. The patient's account of events leading up to the sickness, of sensations felt indicating something is wrong, of interpretations giving the symptoms meaning are all part of the patient's story. The patient's role as narrator in the drama of illness has declined in the twentieth century. This is due to the growing dominance of technologically centered techniques of medical evaluation in which the views of the patient become largely irrelevant, if not obtrusive. The numerical results generated by the analytic technology of the modern medical laboratory, the graphic depictions of illness produced by machines that monitor the electrical actions of the body, the neatly ordered printouts of clinical data developed within modern computers—all these data seem free of the bias and subjective opinion of a person and possess the objective characteristics of evidence we call scientific. Where, then, is there room for the patient's story, told from an admittedly biased viewpoint and delivered to the doctor with all of the inaccuracies, inconsistencies, and omissions characteristic of human memory and speech?

In this two-volume treatise Eric Cassell has set out to fill a large gap in the literature of medicine by giving a systematic account of how language works and how such an understanding can be applied to transforming communications between patient and physician into precise tools of evaluation and therapy. There are few more important tasks for contemporary medicine than this one. In medicine's vigorous pursuit of the biologic knowledge of disease, patient's and their lives have been left out, partly because the method of learning about them—human communication—has seemed an art incapable of describing and thus teaching about with the exactness possible when examining a biochemical event.

Combining information provided by over a thousand hours of tape-recorded conversations between physicians and patients, extensive research into the theories of linguistics, and the experiences of a long career as a clinician, Eric Cassell has formulated a conceptual view of medical communication and a format for its practical clinical application in patient care. He has given form to this subject, and provides a clear agenda of study to enhance our communication skills now and to learn more about them through future research.

It is crucial for modern medicine to establish a balance between understanding general biologic processes that make us ill and understanding the illness as experienced and produced by the patient. To learn of the latter, the verbal and nonverbal elements of human communication in medical care must be understood and mastered. These volumes lead us toward this goal in a more comprehensive way than any other work in the modern literature.

Stanley Joel Reiser

Acknowledgments

This book, and its companion, *Talking With Patients: The Theory of Doctor-Patient Communication,* arose from a study of doctor-patient communication concerned initially with the belief structure of medicine—the system of ideas and values held by patients and their doctors concerning illness, the body, doctoring, and the doctor-patient relationship. Funds for the initial research were generously provided by the Robert Wood Johnson Foundation. As the investigation progressed, it became clear that a practical result of the research would be to provide a firm foundation of theory and concepts on which to base the clinical art of talking with patients—history taking, verbal interaction, and explanation. Once again the Robert Wood Johnson Foundation provided support, but now for the development of a curriculum to teach the theory and skills of medical communication to students and physicians. Margaret Mahoney and Terrance Keenan were extremely helpful in the design of the program. Because of her continued understanding and encouragement, it can be truly said that these volumes would not have come into existence, nor would the research on which they are based been completed, were it not for Margaret Mahoney.

In this aspect of the work as in the earlier parts, both Dr. Bruce Frasier and Dr. Lucienne Skopek were extremely helpful. Robert Mayer and Lachlan Forrow were the research assistants who supervised the recording, cataloging, retrieval, and editing of the tape-recorded materials that are used as examples in this book and in the lectures on which it is based. Everyone familiar with the care, patience, and creativity necessary for this kind of work knows how much I am in their debt.

For overseeing the development of curricular materials and for carrying much of the administrative burden, I was fortunate to be able to rely heavily on Constance Wilkinson Gianniotis who provided endless assistances—editorial, creative, and technical. Dr. Nancy McKenzie, a philosopher, was a member of the laboratory team under a one-year grant from the Blum Foundation. Her clearheaded thinking and questioning always helped show me the way back out of the conceptual hazes in which I would get lost.

In this volume, as in its companion, I have chosen not to footnote. Many physicians, books, and writers have had an influence on my thinking about the clinical process, the nature of the relationship between doctor and patient, and the goals of clinical medicine, and it would be impossible to list them all. However, I would like especially to think Professor Otto Guttentag, a pioneer in the philosophy of medicine, for broadening my horizons in this work. The bibliography contains the names of books that I think the reader will find useful in pursuing these subjects.

In carrying out this project, the focus was always on developing concepts and skills that could be taught and that would be effective in clinical medicine. Therefore I am especially in debt to the many students who have taken these courses over the last several years. Their enthusiasm and involvement has provided both motivation to move forward and specific criticisms on which to base improvements. Over eight hundred patients were kind enough to sign consent forms and wear the microphones that made possible the unique collection of doctor-patient recordings on which this book is based. I am grateful to the many physicians who, knowing that their previously private conversations with patients would be subject to scrutiny, offered themselves and encouraged their patients to participate. I am especially in debt to Dr. Peter Dineen, Dr. Morton Coleman, and Dr. Stanley Birnbaum. From October 1974 to December 1975 the tape recorders were in my office documenting every interaction between me and consenting patients. I must thank my office staff, Esther Hyman, Maria Lopez, and Nancy Levy, for so gracefully putting up with the inconvenience. As always, I am in debt to my patients for their willingness to participate and for tolerating this and the other inconveniences that have allowed their care to be the ultimate test of the ideas in this book.

The National Library of Medicine (Grant LM00056-01) provided the funds that made writing this volume possible. For the tedious and hard work involved in typing transcripts of lectures I wish to thank Nancy Levy. Corinne Standish accomplished the exacting task of transcribing the examples as they appear in the text. The Commonwealth Fund provided partial support for Dawn McGuire, who has been my skillful research and editorial assistant during the completion of the book.

During the time when this book was being written, two of my teachers died to whom I will always be indebted. Cyril Solomon was the first to plant the intellectual seeds that have borne fruit here. I miss him, and I miss the commentary on the manuscript that he might have provided. Walsh McDermott, so helpful throughout my career, understood what I was after and how important it might be to clinical medicine. His support and encouragement was always of tremendous importance, and his death in October 1981 was a great personal loss.

During the past three summers, when the bulk of this book was written, my wife and I have spent August in a small country cabin with two computers and their paraphernalia as our word processors. These pages, as is the case with so many others I have written, testify to my debt to my wife. Sympathetic but honest critic, sure and understanding guide, editor for this volume, Joan Cassell has been the best partner one could wish for.

Talking with Patients

Introduction to Volume 2

Doctors are very powerful people. Properly used, their powers can lift the burden of sickness, relieve suffering, and do enormous good. Used wrongly, their power can do great harm. These truths are part of common knowledge, but most often the power of doctors is associated with their science and technology—with the tools that are employed in the care of the sick. Since doctors have been considered powerful during periods in history when we know that medicine's knowledge of the body and disease was largely nonsense and its therapies virtually useless, their power must reside not only in their technology but in themselves as well. Despite this obvious fact doctors are virtually never taught about their power. Indeed, many become uncomfortable at the word itself, which is a pity because the authority and influence that underlie that power is invested in the role of physicians whether they want it or not. Rather than shying away, it is important to learn where the power of doctors comes from, how to employ it and maintain its effectiveness, what are its dangers and restraints, and the responsibilities it entails.

In the present era we have come to act as though all doctors are equal before the mysteries of disease and in the halls of scientific medicine. In patient care, however, this is not true because kinds of knowledge are required that are not covered by the science of medicine and have not yet been subjected to systematic exploration. Make no mistake about my meaning: medicine would retreat into chaos were it not for science. And no doctor with any sense would even want to go back to simpler technology or lessened effectiveness. This book presumes, however, that scientific doctors who lack developed personal powers are inadequately trained. Another way of wording this is to say that, in addition to the tools placed at their disposal by science, doctors are themselves instruments of patient care who must be refined by knowledge and training to be maximally effective.

It is universally accepted that doctors' education must continue for their lifetime if they are to keep abreast of advances in medicine. It is less well known, but equally important, that a physician continually work at refining the instrument of medical care that is himself or herself, the personal power and effectiveness essential to the very best medical care.

For physicians, in the care of the sick, these personal powers are often called the art of medicine. The art of medicine is composed of abilities in four different but interrelated areas. The first is the ability to acquire and integrate both subjective and objective information to make decisions in the best interests of the patient. The second is the ability to utilize the relationship between doctor and patient for therapeutic ends. The third is employing the knowledge of how sick persons and doctors behave. Finally, the central skill on which all the others depend is effective communication—the subject of this book.

The reader may think that, after the emphasis on the personal power of doctors, the list of the abilities that make up the art of medicine is pretty tame. With increasing experience, however, it will become clear that while one end of the very long spectrum that makes up each of these abilities is anchored in the mundane and obvious, the other end extends to most arcane, and sometimes almost mysterious, aspects of the work of doctors. For example, the seemingly stuffy subject of information acquisition, evaluation, and decision making concerns a difficult but vital skill. Doctors take information about the symptoms and signs of illness, demographics, and the influence of personal characteristics and apply it to sick patients. For the treatment of any but the most trivial illness, however, it is obvious that patients must be considered as complex, changing, psychological, social, and physical beings who are different in different times. To understand them and their illnesses, objective data alone will not suffice, especially since decisions are meant to be in the best interests of the patient and what is in the best interests of one may be deleterious for another. Instead, subjective and value-laden information must enter the decision-making process in equal partnership with objective information. Such a partnership is extremely difficult to achieve, especially since most of us find that "hard" data seems always to win any competition with "soft" data. Yet we know it can be done, because there are doctors who seem able to account in their decision making for the disparate things that make up patients' existences and impinge on their care; such as their beliefs and fears, liver chemistries, family constellation, barium enema findings, hopes for the future, drug eccentricities, immediate needs, cardiac arrhythmias, the kinds of people they are, pain threshold, and the myriad other details of medicine.

It is in the category of subjective information that we begin to approach the borders of our knowledge. For example, students are commonly advised to use their "feelings." But less commonly are they shown precisely what is meant by feelings and how they are to be used. On one occasion a student and I went to the bedside of a man who had suffered so many complications of his diabetes that one could serve a rotating internship caring for him alone. In addition to his other troubles, the patient was swamped by hopelessness. As we left the room the student said, "I really felt hopeless in there." That feeling of hopelessness had been acquired from the patient—the student felt hopeless because the patient felt hopeless. The student was entirely unaware of the origin of *his* feeling of hopelessness. In fact, there were many things that could be done for the patient. In like manner we may feel anxious when our patient is anxious, angry when our patient is angry, and so on. These are the feelings that physicians are meant to use in the decision-making process. But it is obvious that, at least early in their careers, doctors are going to have difficulty deciding when the anger (or anxiety, sexual arousal, hopelessness, sadness, or any other feeling) is theirs and when it was acquired from their patients. To go a step further it seems reasonable to ask how people acquire feelings from others; how they are transmitted? Similarly, where does the information come from that allows one to walk from a room and report, "There was so much tension in the room, you could cut it with a knife." I do not mean to answer these questions, I merely raise them to point out how subtle the arts of medicine can become, even though in their simplest form they seem so obvious.

In the same manner the art of communicating with patients begins with the obvious fact of listening to the patient's words. But at the other end of a spectrum, which includes hearing not only what is said but what is unsaid, of being aware of nonverbal as well as verbal communication, is the use of the spoken language and nonverbal communication as therapeutic tools of enormous power. As with all these arts, learning begins with the simplest aspect of the skill. But no matter how adroit and sophisticated a doctor may become in the use of the spoken language, the basic techniques can never be abandoned or left to chance; they must constantly be honed.

One of the things I hope that you will learn from this book is that it is possible to explore these aspects of doctoring, just as it is possible to explore cardiac function. The arts of medicine have not been subjected to the intense systematic and disciplined study that has been applied to human biology. But there is no reason, apart from custom and habit of mind, why such investigation should not be undertaken. Although the methodologies that must be employed are different from those that are so useful in medical science, research in this area can be fruitful and exciting, especially since so little has been done thus far. Sometimes one discovers things that have not previously been described, while on other occasions one comes to understand the reason why we do what we do. When I teach the material in this book to experienced practitioners, they frequently tell me that they already do many of the things that I describe, but they have never known *why*. It is their experience, as it will be yours, I believe, that knowing the reason something is done makes the skill more effective and more consistently under one's control.

The spoken language is the basic tool of doctor-patient communication, the more one knows about it, the more effective is the tool. A companion volume, *Talking with Patients: The Theory of Doctor-Patient Communication*, describes *how* the spoken language works in medicine: how words do their work and can have meanings and impacts at many different levels, affecting even the body itself; how the attentive listener can know not only what speakers mean but what kind of people they are by their word choice; how all normal speech is logical, and what that knowledge can do for the physician. Readers wishing to enhance their communication skills further will find this information valuable.

This volume is devoted to the actual process of talking with patients: the formal tasks required for the exchange of information between doctor and patient. In these chapters the speaker and listener are considered as an inseparable pair. Here the dyad is always "doctor and patient" (or, "patient and doctor"); there cannot be one without the other. The doctor is not a remote inanimate object talking to a flesh and blood patient. They are both real and exposed to one another—and often both are changed by their interaction. Consequently even when we physicians ask questions, the structure of the questions and their wording provides information about ourselves, our intent, our beliefs about patients and diseases, as well as eliciting such information from patients; "taking a history" is unavoidably and actually an *exchange* of information. This is why I place such stress on the form of questions and the choice of words. The principle that applies is simply stated: the spoken language is our most important diagnostic and therapeutic tool, and we must be as precise in its use as is a surgeon with a scalpel.

Another basic principle will be illustrated in these pages: the illness the patient brings to the physician arises from the interaction between the biological entity that is the disease and the person of the patient, all occurring within a specific context. Here we come upon the true difference between medical science and clinical medicine. For medical science to achieve understanding of the biology of diseases, the disease must be separated from patients: it must be abstracted and generalized. For clinicians to be most effective, on the other hand, we must grasp both abstract pathophysiology as well as how that pathophysiology is modified within a particular sick person. These two kinds of knowledge are equal partners in clinical medicine. To the extent that either mastery of pathophysiology or knowledge about sick persons is slighted, the efficacy of the clinician is diminished.

Understanding that the manner in which an illness develops and presents itself to the patient—and then the doctor—results from the interaction of the particular person (in all the dimensions of that word) with the biological process of disease, aids in focusing the task of taking a history. Too often students are not taught what it is they are after when they take a history. But it makes sense that if one is about to spend an hour asking questions of someone, one should have a pretty good idea of the object of the pursuit. The usual answer, that history taking is done in order to make a diagnosis, is not adequate unless the meaning of "diagnosis" is spelled out. The usual meaning of diagnosis, the identification of a disease, is clearly inadequate to the *clinicians* task. For example, ulcerative colitis, pneumococcal pneumonia, adenocarcinoma of the bowel, myocardial infarction, are all disease diagnoses that are quite specific. But as much information as each one of these names provides, so much more information would be required about each *patient* with these diseases before a clinician could formulate a plan of action. What is the patient's gender and age, how long has the process gone on, has previous treatment been given, does the patient have a long-term trusted doctor or an equally enduring *dis*trust of physicians, is the place of residence rural or urban, what other diseases are present, are a few among many factors that have a bearing on the immediate and long-term treatment decisions. Therefore the emphasis in these chapters goes beyond diagnosing disease. Adding the name of a specific disease should be a late step in diagnostic thinking. Here let me speak of my own practice. What I am always trying to do is to find out what the problem is: what has gone wrong in the patient, and why. As part of this I attempt to determine what I am going to be able to do about what has gone wrong. I think of this process as looking for the place to put the lever so that I can exert maximum force to make things better.

In going through this diagnostic process, I am not particularly concerned whether things are awry in the patient's body or the patient's person. If a patient has chest pain, I am interested in knowing whether the pain arises from the heart, the lungs, the chest wall, or by some other pathophysiologic mechanism; I am equally interested in whether the problem is the pain itself, the patient's loss of function because of impaired organ function, the patient's beliefs about the pain, the patient's perception of self (into which beliefs about the pain might fit), or all of the above. I also want to know why this particular problem arose at this time: is it progressive narrowing of the coronary arteries that finally became sufficiently stenosed to result in ischemia; has blood flow remained the same, but demand increased; did something impinge on a cervical nerve root to result in radiation of pain, perhaps because of heavy lifting or a change in sleeping arrangement; is something injuring the muscles of the chest wall; did the patient's father recently die of a heart attack; does the patient believe that this chest pain represents heart disease and so is avoiding any effort (in which case it is the patient's beliefs that are impairing function); does the patient have overwheming body fears into which this pain fits, although it is not particularly intense; and so on. All of this is another way of saying that in taking a history one should be attempting to discover the process by which a well person became a sick patient in order to devise another process whereby the patient can be returned to maximum possible function. Processes are a series of events that occur over time, whereas diseases are often dealt with as though they were timeless objects—statues in the park. Looking for disease diagnoses, however, has one enormous advantage over the search for the process of illness—diseases are easier to write down, a few words usually suffice. On the other hand, processes are difficult to describe, because a language for process still eludes us. This is why good clinicians usually use disease terms as a shorthand for the process taking place in or around the patient. But one should not be fooled, clinicians write in static terms but base their thinking and action on a dynamic conception of illness.

While taking a history, the doctor should be forming an hypothesis about what is the matter. Because the null hypothesis—the falsification of the hypothesis under consideration—is easier to "prove," since it requires only one solid piece of contradictory information, you should always attempt to "prove yourself wrong." As difficult as it may be, one wants to stop and search for the question that will reveal the *weakness* of one's ideas. A story is being constructed from the facts being elicited that will explain what has happened to the patient. We want it to be an airtight story, so we have constantly to attack it—we would rather have *it* fail then us!

Searching out the clues, putting them together, and testing the premises can be very challenging. When I get on the scent of the process that has made the patient sick, I become single-minded in my pursuit. If I am in an examining room, I begin to pace back and forth as I ask my questions. Nature and the human condition reveal themselves in the operations of sickness, and I find this discovery endlessly exciting and consuming. But the person did not come to see me in order to pique my curiosity or excite my interest, the patient has come to me in order to get better. Therefore no matter how carefully the history has been taken, no matter how thorough the diagnostic evaluation, no matter how certain the diagnosis appears, there is one further question that we must always ask ourselves: *"What if I am wrong?"* The constant possibility of error, and thus of doing terrible harm to someone, is what distinguishes clinical medicine from most other occupations. Thus the diagnostic process cannot end when an hypothesis has been chosen that holds up under test. There must be a concrete plan in case of error. A few years ago I admitted a man to the hospital whom I believed had a pulmonary embolus. The resident disagreed, feeling that pneumonia was more likely. When I returned in the morning, I was dismayed to find that the patient had been treated for pneumonia with no consideration of the possibility of thromboembolism—a possibility that was subsequently confirmed. I was not upset because of the diagnostic error; that happens many times to everyone. The real error was in failing to protect the patient in case the diagnosis was wrong. The object of all diagnostic thought and effort is that the patient be better, not that the physician be proved correct.

When I have gathered sufficient information concerning both pathophysiology and patient, I may in fact be able to make a disease diagnosis. But "arteriosclerotic heart disease," "ischemic heart disease," or "coronary artery disease" (our terms for it have changed in recent years) may be the only disease name entertained. It is also conceivable that the only way diagnosis will enter this case is with the statement "This patient does *not* have ischemic heart disease." Such a statement might comfort the patient; it might even resolve the issue. On the other hand, the patient does not yet know what the matter is, only what the matter is *not*. It is frequently difficult to solve a problem when the only thing one knows is what it is not! Patients are often puzzled, rightly I believe, when they discover that the doctor is not trying to find out what is their problem but instead is attempting only to make a diagnosis of disease. And they are also perplexed, when no disease is found, to be told that nothing is wrong or that the problem is emotional, as if when there is no disease, they are well!

My object here is not to belittle disease diagnosis. If I am sure the patient *does* suffer from ischemic heart disease, then I can bring to bear on this patient's problem all that has been learned by medical science throughout the years about this (reasonably) well-defined concept known as ischemic heart disease. It is useful to remember, however, that for clinicians, making a diagnosis provides a basis for treatment and prognostication and is not an end in itself. This is especially true today, when medicine has become a profession of intervention. For more than a hundred and fifty years since disease concepts, as we know them now, were first formulated, medicine has been primarily a profession of discovery. Medical science discovered things about the body and about diseases, and doctors discovered what disease a patient suffered from. However, until the 1930s there were very few effective treatments; consequently physicians had very little in the way of specific curing actions open to them, and surgery was the only real specialty of effective action. As we are all aware, enormous changes have taken place since then. But, as is so often the case in an era of transition, these changes have not yet altered many basic habits of thinking in medicine. As a result, though we now have fantastic tools for effective action, we are still talking and teaching as though the greatest achievement of a physician is to make a diagnosis of disease! Nonsense. Clinical medicine, with the aid of medical science, has become a profession of effective action. The best thing to do for sick persons is to make them better. To the extent that making a disease diagnosis (naming the disease) aids in that process, it is useful. Much more useful, I believe, is the knowledge of how the body works in health and sickness. When you know what has gone wrong, where function has become malfunction, then you know where to intervene on the patient's behalf.

It is important to realize that intervening on behalf of a patient implies that you knows what is best for that person. With blood pouring from a wound, it is not difficult to know that stopping the flow is in the patient's best interests. Similarly there is little dispute about what actions to take in curing acute pyelonephritis. But although acute situations, such as trauma or infectious diseases, are often used as the model for medical practice, in many situations in modern practice it is not so clear what is in the patient's behalf. Therefore in this book you will find an emphasis on discovering what the patient believes to be the problem, what are the patient's major concerns, and what the patient believes would be a good solution to his or her problem.

In this discussion of taking a history, we have uncovered three basic reasons that suggest why it is so difficult to become a good diagnostician. The first is the usefulness of putting off naming the disease until the last possible moment. This seems counterintuitive; one would expect that naming the disease is what diagnosing is all about. The second reason is that instead of trying to prove yourself correct, you must attempt to defeat your hypotheses. This too goes against the grain. The third reason is that, after all that effort, you should not be concerned with being right but only with making the patient better. This last step, disengaging one's personal and professional pride, is perhaps the most difficult of all. I can remember well an occasion when I *hoped* the patient had carcinoma of the lung because that was the diagnosis I had made, and I was appalled at myself for feeling that way. It has happened more than once to me, and I am aware that it happens to other physicians—perhaps all. But as the years go on, the competitiveness can be tamed or turned to more productive uses. Until such personal traits can be subdued, diagnostic excellence remains to be achieved. Since controlling competitiveness and vanity seem to be tasks of a lifetime, the goals suggested here require constant effort. I believe that for physicians, at least, the game is worth the candle.

The examples that form the basis for the text were drawn from more than a thousand hours of tape recordings of naturally occurring doctor-patient interactions. Over eight hundred separate patients (not just visits) are represented. The recordings, the bulk of which were made in 1974 and 1975, come from private offices, hospital rounds, and to a far lesser degree from clinics. I recorded every interaction between myself and consenting patients in my office and the hospital for more than a year. My research staff then chose segments that seemed best to illustrate the points being made and whose sound quality (which was generally very good) would permit them to be used as recorded examples. They were then edited for clarity, with every attempt made not to alter the essence of the transaction by the process of editing. The method of transcription, chosen from among the many that were tried, approximates conversation as much as possible without being impossible to read. Except where indicated, the roughness and redundancy of the spoken language have not been smoothed—the "warts" have been left in place.

The examples, besides illustrating issues in communication, also involve actual cases and approaches to the care of patients. The reader will become closely acquainted with how I practice medicine, what I believe the nature of the relationship between patient and physician should and should not be, and even how I treat certain illnesses. In addition the personal approaches of other physicians are exemplified.

When I first started listening to recordings of other physicians in their offices, I was pleased to hear that the same things, some of them quite strange, happened in their offices as in my own. Doctors who have listened to me with my patients have expressed the same sentiment. The practice of medicine is a very private matter. The most intimate aspects of a patient's life are revealed in physicians' offices, ranging from what kind of underwear is worn (or not) to what the person secretly thinks about other family members, as well as the overtly sexual matters that are usually associated with the word "intimate." At times the doctor is as exposed as the patient; for this reason I admire the doctors who put aside their reservations to wear my microphone. Because of this readers should understand that they are privy to information shared by few in the past. For the same reason and also because it has a single author, this book will be unavoidably idiosyncratic. Despite the drawbacks of this personal quality I hope you will find learning about how another doctor works as interesting as I do.

Because this approach is neither quantified nor treated statistically, it may cause discomfort to physicians who have been raised on numerical "data" and taught to avoid the "anectdotal." These readers should be aware, as the philosopher of science, Alfred North Whitehead, has made clear, that it is impossible for a methodology to discover or demonstrate something that does not exist within the system of ideas on which the methods are based. So it is with the personal and subjective qualities of patients, which are illustrated in this book and which play such an important part in medicine. The fundamental cannons of modern science do not recognize issues of subjectivity and values, so we are forced to use other ways, besides quantification and statistics, for demonstrating and teaching these aspects of patient care. In fact I believe that there is *no* other equally effective or realistic manner to approach the study and teaching of communication between doctor and patient in the clinical setting than the use of examples. Indeed, the recitation of cases—telling stories—has been a method of teaching aspects of clinical medicine that has survived through the ages because nothing else does the job as well. Recently scholars have begun to direct attention to the stories about patients that are used in teaching because they carry a kind of information that can be transmitted in no other way.

It is appropriate that a book about how the spoken language is used to know what has gone wrong, and how to make it better, should start with a patient's story.

1

"Tell Me the Story of This Illness, Please"

In this chapter I think you will learn that a patient's history is more complicated than it appears on the surface. It has diagnostic value because of the interaction between the physician and the patient and because of what the doctor brings to the interactive process. In this chapter the "history of the illness" will be dissected, to show what it is made of and to provide a basis for later chapters demonstrating "how to take a history."

Most students, even most physicians, have never heard someone else take a history from beginning to end. It is sad that even during the period when this crucial technique is being learned, most instructors never listen to their students taking an entire history. The reason usually given for this lapse in teaching is, of course, that the process takes so long. Can you imagine a surgeon saying that surgeons in training are not supervised for a whole operation because it takes too long? Instead of being certain that students fully understand how to carry out this process, one that will be basic to the care of patients for the rest of their lives, we give a couple of "how-to" lectures and let it go at that.

In this chapter you will read many histories and find out what they have in common. The histories appear verbatim, except for two changes. The first is that they have been edited to remove all, or nearly all, the doctors' questions. What is left is what the patient said before any questions were asked, or what a history would be like if no questions were asked. The second change, for this chapter, is that representations of paralanguage have been removed for ease of reading: the pauses, "ums," and "uhs" are gone.

Here are the patients' own words. The first patient is a fifty-four-year-old woman, with thrombophlebitis and pulmonary embolus, who was admitted to the hospital as an emergency following this, her first office visit.

All right, tell me the story of this illness, if you would, please.

Well, it started a few weeks ago, I was visiting my sister in Rochester. I wasn't at home. I was far from home. The first thing that happened was that my eyes swelled up. I'm telling you everything in chronological order. They got kind of black and blue. Now, this has happened before. It's happened about four times since the beginning of this year. But it seems to be, or so I was told, a contact dermatitis, and possibly unrelated to anything that followed.

Anyway, my sister took me to a friend of hers up there and they seemed to feel that it may have been something I touched or wore and they gave me some Lidex Cream to put on my eye, and then, you know, that hasn't returned. However, . . .

But before I was leaving, I think the very next day, still there, I was just sitting at the table and suddenly I got a very bad pain the back of my left leg. About over here. As if maybe a muscle were pulled, or something like that—just ached. A Charlie horse. And then that developed so that within two days I could barely stand on that leg. I thought for a moment maybe it's something vascular, but my other leg is the bad leg, from the vascular point of view. I had thrombophlebitis in it. That was my good leg.

Anyway, that lasted for about two days and then if I stepped on a stair going up or down, it was very painful. And then that started to subside. But before that subsided, I started getting a sense of pain in two other places. One was abdominally—not cramps really but just a pain, and another was in the middle of the back. I don't remember which one came first, but the whole procedure has been one thing following another.

I tell you—one thing about the pain. Sometimes I get a spasm, apropos of nothing and it just goes right through. And then last night, for example, I woke up and there's a spasm, but it seems to be vibrating. The only way I can contain it is by getting into a tub. It's constricting and—

Right.

And then I had a feeling that it's going, moving, maybe moving in the direction of the heart.

Right.

And that I—

Isn't it fearful?

Well, ye— I'm here because last night I really felt fearful.

Mm-hm.

What is her chief complaint? A problem with her eyes? Pain in her leg? Chest pain? If you choose chest pain, is it because it is your idea of her chief complaint, or hers? In fact the patient did not give a "chief complaint"; she presented a sequence of events. She is correct when she says, "The whole procedure has been one thing following another."

All histories of illness are stories, and this patient told the physician a story. All stories have certain things in common: they are a series of events that happen to some protagonist in some place over a period of time. Most stories also have reasons: the reason the events described took place. Stories, like newspaper reports, have a who, what, where, when, and why. Here is another story.

Now, tell me the story again, please?
Wednesday evening, we were up at a nice zoo in Tacoma.
Right.
And we went out to a place to have lunch.
Right.
It's a crazy sandwich place; some kids were running it. And we had lunch, and about three o'clock, two hours after lunch I felt—not cramps exactly but somewhat gripping. Ok, I've had a stomach ache for an hour or two before so we walked around and finally I told Judy I wasn't feeling well, so we go back to the motel. So we go back there then—I'd gone to the bathroom a number of times—two or three times that day. Fine, no problem. And then it seemed to me that the gripping shifted. It wasn't every minute. It'd go on for a minute or two and then off, minute or two, then it shifted to a lower part.

And the— so I took Super Pectin, a couple teaspoons— tablespoons full, and some Titralac and a couple of Tylenols, 'cause my head was—I guess I was getting scared. And I was very uncomfortable, so we had a heating pad, I put the heating pad on. I got about five to six hours sleep Wednesday night. I had to sleep on my stomach, though. It was the only position I could get comfortable.
Right.
And then, in the morning it persisted. The same thing. A few minutes, then it'll ease up—but I told you, whatever it was, it worked its way down. But it seems to be centered around here, in the middle of the stomach.
Right.
And it didn't work its way down. So Thursday morning I spoke to Judy, who said, "The hell with this," and we caught a flight and, in about eight hours, last night arrived in New York. And this morning, I went to the bathroom just before I came here. It isn't as bad, but it still occurred even when I was driving in and even for a few seconds in the office.

Who are the characters in this story? The narrator is one character. Judy is another. There are the "kids who ran the sandwich place."

But there is another, much more important character. There is an "it." The pain is part of the "it," and so are the other symptoms. "Symptom" is the physician's name for events that take place because of that character, the "it." The other unspoken character in this story, to which things are also happening, is the narrator's body! This is a medical story, an illness story, and illness stories are different from other stories because they almost always have at least *two* characters to whom things happen. They always have at least a *person* and that person's *body*. It is very important to make the distinction between the person and the person's body as you hear the history of an illness. It may be surprising that I seem to be insisting on dividing the person from the body, because I have placed such emphasis on listening to the patient's personhood and on the contribution of that personhood to the patient's illness. In fact I do not believe in separating person from body unless it serves a purpose. In history taking, however, it is essential to distinguish what is occurring in the body from what is happening to the patient, in order to serve the interests of both.

This next example displays both the person and the body very distinctly, because the woman is very much a person, and her body is doing some very contrary things. (We saw this example in volume 1.)

Two weeks ago, I'm sitting in my office and I start getting these spots in front of my eyes. I couldn't see anything. I was— you know, nothing. I was just—seeing these little, crazy little spots, which— I've had that before, but maybe for a minute, or not even a minute. A couple of seconds or something, I'll see spots.

So I said, "Well, you know, I'll stand up, I'll walk around, it'll go away. I'll take out my lenses." I figured, "It's the lens— " I wear soft lenses. So I popped out the soft lenses, and the spots didn't go away so I took a walk and I got as far as the Art Department and I couldn't— well, for one thing I was blind without my lenses and I'm sort of walking like this.

But really it wasn't just the spots, it was a combination of things. At that point I didn't really have a headache, I don't think, I just felt slightly dizzy or something. It was a couple of weeks ago, I don't remember exactly. But I sat down in there and I was figuring, "Well, what is this?" And I started getting these shooting pains in this side. It was just like all—everywhere. And it just made me a little bit upset. I figured, "Well, I didn't have anything to eat too much this morning. I'll go across the street to the Beanery, you know, and get something to eat."

So I got up from the Art Department and I went down and I was standing on the corner of 52nd Street and Madison Avenue and all of a sudden I just— I guess 'cause I was worried because nothing like this really, I don't think, ever happened to me—well, I know it didn't—shooting pains in my head and because I was scared about that, I think this is why this happened, but I felt compelled to hold on to a tel— a metal pole thing, like this, 'cause I started feeling woozy. And then I figured, "Ah, I'm going to get into that place and have some orange juice, and high protein stuff, I'll feel better."

Then the light changed and I jus— I don't know what happened. Just sort of, my knees went out from under me. I just fell down and I was carrying my coat, and my coat dropped, so I just went down like this, I got my coat, and I stood up, you know, so that it (I guess) looked to people like I dropped my coat and I just— you know. But really that isn't what happened. I fell down first.

I go into the place and I say to the guy, "Give me this, this, and this." And I ordered some stuff and I'm waiting for it— No, I'm sitting there and th— th— it's sort of getting kind of worse and all of a sudden—and I— I'm— I'm trying to keep sort of in touch. Like I'm saying, "Well, this is all psychosomatic, or something like that. Nothing bad is going to happen." And then I just sort of freaked out just for a little tiny bit and I just had this horrible image— I had this image that my brains were like bleeding or something. It was very strange. And I just figured, like, "Ah! This is curtains. I'm going to croak. I'm not going to get out of this place!" And I kept thinking, "Isn't it funny how, I'm sure that people must die of embarrassment. Because I'm in a drugstore, I could have gone—if I was really that scared, I thought I was going to die. I figured, "This must be a hemorrhage or something," and started thinking of all these things. And I didn't want to bother the guy in the back to say, "Well, call my doctor," or something.

And then a little bit after that is— is that I got this headache, like a regular headache, like right in the front, like I've had before and it was terrible. But I took some aspirins and stuff, but I was— because I can deal with that. I know what that was. And then it wasn't until I was OK, I guess. The headache finally went away. It took a couple of hours but the other thing was fine. And I told a friend about this and she said, "Oh! Well, for Christ sake, that was a migraine!" And I thought— I said, "Ah!" (Snap) I mean I didn't— if I had thought of it when it happened, I probably would have felt much better.

This story is very clearly told. In one form or another, something like this has happened to all of us. A set of symptoms occur that are very frightening, and then something else happens that clears up the mystery and removes the fear, all in a moment. She is such an articulate patient that we find out not only what happens to her and what she does about it but also her causal thinking. She even describes that most fascinating phenomenon of sick people, that they can "die of embarrassment." Instead of calling for help, not wanting to bother anyone, people will sit in their car or office and have a heart attack or stroke! She describes these two sets of events—the things that happen to her and the things that happen to her body (her symptoms)—in a manner that mixes them together. She does not make the distinction between her person and her body, as does the man in the previous example. But it is not necessary that the patient separate body from person in telling the story, only that we do so in hearing it.

Would you have made the diagnosis of migraine headache from the story as told? If you did, you probably have migraine headaches. Do you know what was the matter with the man who returned home from Tacoma? As you probably surmised, he had gastroenteritis. Do we know that because of what he said, or what we (without awareness of our doing it) added to his story as it was told? He gave such sparse information concerning his symptoms that many other diseases could be compatible with those he did describe. If you believe he had food poisoning, it is probably because he mentioned the "crazy sandwich place run by some kids." In other words, he gave very little concrete evidence on which to base our diagnostic conclusions, providing us instead with *his* diagnosis, his beliefs about what was the matter and why.

At this point it is quite reasonable to ask what it is that we go after when we make a diagnosis. Is a disease an object like, say, a car? What is gastroenteritis or thromboembolic disease? If I tell you that I have a patient with carcinoma of the breast, you have been told little in terms of taking action. "I have a patient with scirrhous carcinoma of the breast. What should I do?" Before responding, you will need to know the age of the patient, stage of the disease (indeed, a patient may be said to have carcinoma of the breast who had a mastectomy three years earlier and now has no signs of disease), operated or not, size of tumor, number of lymph nodes involved, hormone receptor status, previous treatment, whether the patient lives in a remote area or near a medical center, how much the patient knows about the disease, who I am (surgeon, internist, or oncologist), and more. You cannot even assume the patient is female!

A disease is not a thing, it is a process. A process is characterized by change over time. Events are steps in a process and can be identified in space and time. Just as in a story, events occur over time, so they do in a disease. In other words, not only is the history of an illness a story of events happening to characters through time, but a disease is a story, too. Of course the special name for the story of a disease is its pathophysiology—the unfolding of an abnormal set of events in the body.

It might go like this. A person injures the left lower leg, by striking it against something hard. Almost immediately it becomes painful and ecchymotic, but the pain soon subsides. About ten days later, another pain develops in the same calf that is different from the original pain, which had by now gone away. The new pain is more achey and sore. Looking down at the leg a day or so later, the person notices that where the ecchymosis had been, there is a place now more swollen and reddish and warm to the touch. After a few more days the ankle is seen to be a bit swollen. Perhaps four days later, while riding in a car, the person gets excruciating pain in the left side of the chest, a little lower than the nipple, which makes breathing very difficult. Within a few more hours the pain is much less severe. But shortness of breath is more prominent, and some staccato coughing accompanies it, occasionally productive of some darkish bloody sputum. By the time the person gets to the emergency room, the pain is gone. This is the story of thromboembolic disease following an injury. We know, step by step what happened in the body to produce that story. The injury to the vessel wall was followed by the thrombosis, followed by inflammation in the same area, followed by some venous obstruction, after which a piece of thrombus embolized to the lung where infarction took place (the bloody sputum), producing a pleural reaction (thus the pain); as pleural fluid accumulates, the pain goes away. Medical scientists know the story at other levels: macromolecular—involving the clotting process; micromolecular—involving, say, the platelets; the organ level—involving blood vessels and lung. Although we are intensely interested in these other levels, because they provide the basis for our understanding and thus for our actions, we are, at the moment of taking a history, involved only in the whole organism level: *How each of the things that happened at every other biological level is reflected in the way the person describes his or her functioning as a whole organism, and that is the only level to which we have access while taking a history.* It is a common error for physicians to become confused about which level of human biology we are operating on, at any one time. To reiterate, when taking a history, we are operating only on the whole organism level and are interested only in how malfunctions at any other level are reflected in the functioning of the whole human being! What each physician who cares for patients must develop is the ability to know *how* dysfunction at other levels—whether it be molecular, cellular, or organ systemic—is expressed at the whole human being level, and just as important, we must learn what a malfunction at the whole human level tells about the state of other levels.

To make the relationship of illness history to pathophysiology even clearer, we should see that there are levels above the whole human that are involved in the story of an illness. It could be that the story of the man from Tacoma starts even before he gets to the restaurant. Pete and Joe are a couple of the "kids" who run the "crazy sandwich place" in Tacoma. Say, few days before Judy and Jerry went to the restaurant, Pete cut his finger, which became infected. The morning of that day, Pete was making egg salad for the egg salad sandwiches, and Joe told him to put it in the refrigerator and come help out front. Pete forgot to do so (actually he resented Joe always bossing him around), and when Joe returned just before lunch (about five hours later) and saw the egg salad on the counter, he figured Pete had just removed it from the refrigerator, in anticipation of the luncheon rush. Now, Judy and Jerry enter the restaurant and have egg salad sandwiches on whole wheat, and coffee. Six hours later the story of Jerry's illness starts. These details concerning how staphylococcal food poison is spread are intensely important to not only to epidemiologists, who must concern themselves with social organization and interpersonal relationships, but also to physicians, for whom they provide understanding of the wider contexts of human illness and possible areas in which to take action. But for the physician taking a history the levels of biological systems that are higher than the whole human are only of concern to the degree that they are expressed at the whole organism level.

I do not want to give the impression that patients tell stories of their illnesses that can be translated directly into pathophysiology. If that were true, the premise of this chapter, that the patient's history of illness as given is frequently valueless, would be incorrect. Let us consider another series of histories.

So what can I do for you?
I've had a growth on my back for quite a few years. I've had surgery quite a few times. So I've had it looked at each time. I was asked whether or not it was getting any bigger, the answer was "No." I was asked if I have any pain—the answer was "No."

This last month and a half or so it's gotten much larger. Then I went swimming and I attribute the itching that I got to either the salt water or chlorine. And then I got a bop on it, accidentally, and now I do have pain. But I had the pain before I got the hit. I can tell this time I had no pain.
Let's take a look and see.

This man had an infected sebaceous cyst. His description was less than graphic.

Here is another.

So how are you feeling, Mrs.——?
Well, actually a couple of weeks ago—
Yeah, what happened?
Well, I ran into a bad case of diverticulitis. I know I've had it before.
But when I was here for the different tests a year ago, I had been well
and without it so long that it slipped my mind—I really didn't call
up for that. I called up for some other— Anyway, I was forced to go
to a doctor across the street because my temperature was 102. So I
couldn't wait, Doctor, I had the appointment because I was too bad.
So now, I'm still a little bit on the shaky side. I am better and I
brought the medicine that he gave me with me to show to you. And
that's all I can tell you.

What was this woman's problem? If you believe that she had
diverticulitis, it is only because she said so. As is often the case, she
gave a diagnosis, not the history of the illness. We are not seeing the
clear-cut progression of disease through the body discussed earlier.

This next, however, is a longer story that *is* quite cohesive.

All right, tell me the story of these headaches, please.
See, this happens in April, last year. They start to— In the beginning
was— I felt like little blindness in this right eye. And this lasts about,
let's say, ten minutes. And, you know, I thought, "I have something
in my eye." You know, when something cover you eye? I try to clean
my eye but that's, you know, stopped for about ten minutes.
  Then the second time that happened was about a month, a month
and a half, you know, in the park, that I had the same thing, but this
time was a little longer and I said, "Ah, this is nothing." But I waited,
and it was longer than the first time.
  And then happens again in about, let's say, maybe a month later
again. This time, you know, I was in the office by myself, and I did
have a sign on the door. And I tried to read the sign on the door, and
complete. This eye cannot see anything. And that time I can never
read, like, I could see "health service," but "student," that wasn't
right—you know, the left side can never read that word. So I try, and
I say, "Let me see what happens." I try and I go, "Cannot read!"
  But I don't want to say anything to nobody. I just see how long this
going to, you know, last. So it happens that a student came in and I
have to look for her chart and I had to look under her name. And then
I went back to the filing room and I try to file some paper. It was very
hard, very difficult for me to read the names! And I keep trying to test
myself. I say, "Let me see," and then I try, I cover one eye and I go—
and then I decide, "Oh my God, this—"
  But I told one of the messengers that came in, I say, "Listen, I have
completely blind in this eye for a half hour!" And he told me, "Well,
you know, talk to Pat." You know. And I say, "Let's wait to see how
long this happen." I don't want to tell her because she will see and

send me to the Emergency Room, and I don't want to do that. So I just say, "Now, let me wait."

But about an hour I start seeing—you know, my vision came like, little by little. And then I went back to her, and I explained to her what happened to me. She say, "Marie, why you don't tell me?" So I say, "Nah, I shouldn't tell you that." You know, and she decided to send me to the Emergency Room. And I say, "No, let me go to the clinic." And I explained this to the doctor.

So I went to the clinic, and they gave me an appointment for eye test. And I went through all the eye tests and that blinding doesn't come to me any more, and I see Neurology, you know. And all this step that I did. This doctor that saw me in Bellevue recommended Dr. H—— at NYU. And I explained, you know, the way that happens, and I ask him, "Do you need any X ray from me to find out what's wrong?" He say, "I don't want to 'spose you in these X rays." And "I think it's migraine headaches." He told me and well you should—

Wait a minute. I don't hear headaches here. When did headaches start?

She never mentioned the headaches! The patient tells a long story about headache but says nothing about about headache. Instead, she describes episodes of blindness which are quite impressive. One might easily explain the patient's leaving out the headaches because blindness was the frightening symptom. What is important to us, at this moment, is not why patients do what they do, but what the phenomenon itself looks like.

In this next example, the object is to see if you can figure out where the ache is.

OK, go ahead.

See, a week ago, Sunday, I fell on the street. And you know, I fell like this. I went—like that. And this, my arm, I hit right here. You know, I felt where it is. And it aches here. And it ached, a little, then a couple of days it seemed to be all right. But now I know, you know, I know that every night when I go to bed, when I start to move around, I feel an ache in there. Sometimes if I cough. (Cough) Well, naturally it doesn't pain me now. Just now and again. But I noticed that if I'm lying on my back, then when I move, it aches. It's not a terrible ache, but it makes me very nervous.

It makes her nervous because the ache is in her breast! Once again, apparent vagueness about something so important to the patient, might be explained by saying that perhaps she pointed to the ache while telling her story. Perhaps, but perhaps not. Let us continue with the observation, leaving interpretation aside. Remember, however, that these histories are natural, except that the doctors' questions have been removed. In most examples, what you are reading are the opening moments, when the patient begins to speak. When I know where the ache was, or that the patient had a sebaceous cyst, it is because this information was revealed later on in these interactions.

These histories come from patients in doctors' offices. This is an important distinction because the hospitalized patient, particularly in a teaching hospital, may have been "trained" to give the history in the way that the physician prefers to hear it. The patient may emphasize what doctors seem to find important rather than what the patient thinks is significant. Thus the hospitalized patient's history usually sounds more organized and to the point than the stories presented here. But, for the hospitalized patient or for the outpatient, the same truth holds: the history as spontaneously provided is not much use; after all, how does one know whether the doctors who "trained" the patient to give a good history were themselves correct in what they believed to be the important part of the story?

In these examples the patients are not objective observers, reporting on the march of disease through their bodies. They are, instead, telling about things that happened to them. We do not simply hear the facts from these patients; we hear the facts along with their explanations, interpretations, and understandings. In volume 1 I discussed at length the way in which every event must be assigned meaning—this is part of the human condition. Here, we are dealing with the problems caused by this characteristic of the mental life. Sickness is a series of things that happen to someone. That person *must* assign meanings to these happenings, and the meanings must have coherence for the sick person. Doctors are no different in this respect (which is why it is so difficult to learn to separate observation from interpretation); we too must assign meanings to the events the patients are reporting. It is not the patients' fault that doctors often provide interpretations or understandings that are different from the patients'. In the vast majority of instances patients are trying as hard as possible to be cooperative; after all, they came to the doctor in order to get better. As anyone who has ever gone to a physician can recollect, people try to decide what to say first and how to tell the doctor what is the matter. They really wish to be understood, and the last thing they desire is that the doctor make a mistake because the patient told him or her the wrong thing! There is also the status issue: after all, the patient is going to see an important person, *the doctor*. The patient does not want to appear dumb, ignorant, "neurotic," hypochondriacal, or as wasting the doctor's time. ("I hate to bother you, I know how busy you are, but . . .") Thus many patients plan what they are going to say.

It was my original purpose to inundate you with examples of patients' histories. Therefore let us have a number of them in a row.

So, how've you been?

I've been well up until—I would say last Thursday. Then on Friday I was very fatigued and not well, and I'm home from work because of that reason.

So even though I rested over the weekend, it didn't seem to be sufficient to do the trick. I have been working, and I know we've had very bad weather—and we had the blackout, you know, and all the things that would contribute. But I don't feel that it's that. I don't know that it's my pressure. The trouble is with my head—dizziness, light-headed, then alternately heaviness in the head and an all-over feeling of pretty tired.

How do you feel now? Do you feel— .

Well, I— I feel all— yesterday— yesterday—

Go ahead—

Yesterday I was getting ready— I went to church. And I was going to my sister's and then my right eye— my right eye— my right eye, was OK. But all of a sudden— of course, this happened before, long ago, that all of a sudden I saw something. It was a light, but like this—and I said, "Oh, my God."

So I was wondering, do you think I could see an eye doctor or something? Wh— I don't know—

THIS is what I called up for—I've got a funny rash on my legs. And I got down in the basement today, going down and feeding a lot of kittens. And where they had— and when I went down there, there was a million and one fleas on me!

So what do I do. I spray my legs with flea spray. And I don't know whether the flea spray did it or what. It was so many of them there that I had white socks on, they were black—

You should—

I don't know what happened—I didn't know that they were there.

All right, what can I do for you?

Second of September, I was hospitalized in— well, I went into French Hospital because I had had a history of blood clots and I didn't like the way my left leg was feeling. So I went over. It was a Tuesday night, I knew he was off Wednesday, so I go to the Emergency Room!

Well, they kept me, and they started anticoagulating me. And at any rate, after five days I was having a lot of problems at French Hospital, pretty much with the nursing staff.

When was this?

September 2nd. I was having an awful lot of difficulty. And I left. I left the hospital and Dr. O—— put me into Park West, where they continued to treat me for thrombophlebitis in my left leg.

But, the thing is, is that it's not feeling better. And they could not keep me on the anticoagulators because my blood level kept going all the way down, and I'd start to either bleed or bruise. Anyhow, the biggest main complaint: I have been running a low-grade fever for a while now, and the leg feels rotten. It's the muscles of the leg. If I can describe to you the way it feels, it feels like the muscles, both my thigh and my calf and it even goes down into my foot sometimes, it feels— the muscles feel strange. They hurt, but they also feel peculiar. They feel as if they're spasming.

Now, how did it— I mean, when was the fast pulse first noticed?

Well, it seemed—to start all at once. About a year ago. I'm sorry I don't have the records. About a year ago, I was bowling when all of a sudden everything went dark and came back. I didn't— didn't pass out but it's the nearest thing to it, I guess. And it scared the hell out of me. And it all started— it seems like it all started at once. Maybe it didn't, but at least that's when I became aware of it, and also it was soon after that that we decided that part of this was due to psychological pressures and all that kind of thing. And I started going to Dr. S—— too, during this period.

Again, in these examples, not much of solid diagnostic worth is presented, because what has been described is what the symptoms mean to the patients more than the symptoms themselves. If this occurs frequently, which indeed it does, it might help us to examine the way in which patients' actual histories differ from the mere reporting of symptoms.

A good place to start is by defining what a symptom is and what it is not. A symptom is the patient's report of what is experienced as an alien body sensation. The key point is that the symptom is *experienced as* alien or unusual. Not all symptoms are abnormalities, and not all abnormalities are symptoms. Sometimes I hear a doctor say "That patient says it hurts, but I don't think it does." If the patient says it hurts, then it hurts! Otherwise, the patient is lying, which is not usual. What the patient means by "hurts," what that particular "hurt" means to the patient, and what both mean to the doctor are all separate issues. *If the patient says it, then it must be presumed to be true.* What the "it" is, is the question. The opposite problem also exists. It is a common experience to hear patients report that they do not have a cough and then hear them coughing during the examination. When queried, they may tell you that is not a cough, it is just their cigarette cough! "Cigarette cough" has become part of them, like age wrinkles. I remember a woman who had rheumatic fever as a young child and had been hospitalized many times before adolescence because of her heart disease. It was clear to me, from her history and examination, that she was going into congestive heart failure; however, no matter how I questioned her, she denied that she was short of breath. My suspicions about heart failure and her denial of shortness of breath were contradictory. Then I started to ask her in great detail about her day. From the moment she arose, she would do a little and then rest, do some more and rest some more; this was the way her day went. Her dyspnea was not a "symptom"; it was a way of life. Two things are to be learned from this anecdote. The first point is obvious: no matter how "abnormal" something is, if the person has become acclimated to it, the abnormality may not be reported as a symptom. The second point, to be considered in greater detail in the next chapter, is that symptom names—shortness of breath, for example—are shortcut labels for something the patient experiences under certain circumstances. If questions that use the label (shortness of breath) do not evoke a response, then questions designed to help the person relive or remember the experience may be necessary. It should be noted that the reason this patient said that she was not short of breath may be more complex than merely habituation to a symptom. Denial may also have been operating. This is not the simple denial of the patient "forgetting" the symptom but rather an alteration in life-style, so that the symptom is totally avoided. In the similar manner patients with angina pectoris will sometimes change their habits so that they *never* exert themselves sufficiently to bring on their chest discomfort, because they are so distressed by the symptom and/or its meanings.

The key word that describes symptoms is *experiences*. A symptom is something the person experiences. In every facet of our existence, and in every minute of our lives, events are constantly impinging on us. The entire sensory apparatus, from the proprioceptors to the eyes, is collecting information at all times. By whatever method the input is monitored, change is instantly recognized and often acted on. The recording of these experiences seems to go on below awareness most of the time. (How chaotic it would be if we were aware of all the things that happened in and on the perimeter of the body!) In some manner, however, much of this information is available for recall when necessary. In hypnosis, subjects can be made aware of sensory information from within the body as well as outside. They can often be shown to have recall for such information, even when they could not consciously call it to mind: smell, taste, visual images, the feeling of something against the skin, abdominal sensations. One does not need sophisticated methods to demonstrate this point. You have parked your car and walked down the block; then your hand tells you that it did not lock the car. When you go back, indeed the car is not locked. I am not competent to discuss the mechanism of memory, but I do wish to remind you that in such circumstances, it is a feeling in your hand you turn to as a source of information. In the simplest terms your hand experienced sensations that were available for recall. Like the unusual "sense of absence" in the hand, symptoms are a change from everyday sensations, and they seem to produce a similar nagging awareness of their presence. Evidence concerning pain perception suggests that there are thresholds of awareness for which individual differences can be demonstrated. It has also been asserted that there are ethnic differences in pain thresholds (Zborowski 1969). When the stimulus has reached awareness, the person is presented with an experience that must be given meaning.

In an earlier work I suggested that, in addition to the *experiencer*, the person also has an *assigner of understandings* (Cassell 1979). The report of a symptom, I suggested, was like a sentence. In a sentence the speaker acts as a coupler between what is said and what it is said about (Percy 1975). In the same manner a patient reporting an experience is a coupler between the experience and the verbal report of the experience. As we have already seen, to articulate the report is to assign meaning, and to assign meaning is to assert the speaker's understanding of the experience.

As you read these next paragraphs, three things will become more clear. First, the facts of the experiences, the body sensations that provide the basis for symptoms, become altered as meaning is assigned. Second, the manner in which the experiences are altered in the reporting tells much about the patient. Third, the information obtained in this manner can be used to aid in treatment. (This third point is covered more fully in chapter 5, Information as a Therapeutic Tool.)

The Assignment of Value

In volume 1 I discussed at length how the word choice of speakers not only describes their experiences but also suggests how they feel about them. When you listen to patients relate their history of illness, this knowledge becomes most important. We learned that adjectives, verbs, nouns, adverbs, and pronouns all tell the careful listener about speakers' attitudes toward the objects of their sentences. And in the chapter on paralanguage, it was demonstrated how the nonword portions of utterances—pause, pitch, speech rate, and intensity— also contribute to our knowledge of the speaker's beliefs and values. It follows that it is virtually impossible for a patient to report a symptom as an "objective" fact, as would a talking computer. Adjectives do not merely modify nouns, such as pain, nausea, dizziness, or swelling; they endow these experiences with value relative to the patient's other values. As we have seen, a pain may be described as mild or excruciating, uncomfortable or sore. The assignment of value can be both idiosyncratic and shared. Certain kinds of pain—pressure on a nerve root for example—will be described by the same kinds of adjectives by most people. Such commonality allows adjectives to enter the diagnostic process, much like an objective referent.

Because value assignment is a part of all utterances, a trained listener can hear the scale of values that the speaker generally uses. This enables the attentive listener to understand the patient better, but it also permits the doctor to calibrate in his or her own terms the experiences being reported.

The Experience of Others

Patients' reports of symptoms are influenced by their association with the experiences of significant others—primarily family members. The importance to the diagnostic and therapeutic process of these human connections will be the subject of chapter 4. At this point, however, let me state unequivocally that what the patient knows of illness and health through the experiences of the patient's family, is far more influential in the assignment of meaning to alien body experiences than other, more impersonal, sources of knowledge. These beliefs cannot be disregarded or brushed aside without risking significant errors in history taking, as well as in every other aspect of talking with patients.

The Space-Time Dimension of Experience

People seem to vary enormously in their ability to report details about the time and place (including the place on the body) of experiences. In part, this appears to be idiosyncratic—some patients remembering in great detail events of the distant past and some assigning everything before yesterday to the hazy past. It is my impression that the future may be handled by different individuals in a manner similar to their handling of the past. Time and place are often related:

"It must have been 1975 because that was the summer we were in New Hampshire." Time may be used as a mechanism of denial, so that the events become difficult to recall, relative to the distress they bring to mind. In the opposite way, "I remember it as though it were yesterday" becomes a means of maintaining the largeness of an event. Terribly threatening events that are remembered seem to remain, for most people, close in time. "That accident couldn't have been a year ago already!" Similarly a bad prognosis, even though far in the future, may be dealt with as though the threat were imminent. For example, one patient with rheumatic valvular disease, when told that an operation might be necessary in two or more years, began behaving as though the surgery were imminent; she lived in constant dread. For another patient, saying that the medication must be taken now to forestall some far future event (as in the prevention of stroke or heart disease in hypertension) is the equivalent of diminishing any sense of the importance of the event and, therefore, of taking the medication. In all of this, time seems to be managed as though it were a spatial, not a temporal, dimension. The more important the event is considered, the closer to the present self it is conceived.

How does this dimension influence history taking? A patient with a right-sided abdominal mass and fever was considered to have an ameboma because his symptoms started while on an Asian journey. In fact, however, the first episode of abdominal pain had started a month earlier, in London, but was attributed by the patient to carrying a heavy suitcase. Careful questioning that related times and places on the trip to body sensations (distinguished by the history taker from the "causes" for them that had been assigned by the patient) produced the classic story of an appendiceal abcess: the original appendicitis, the relief of pain following rupture, a few days of fever, then quiescence, followed by a return of fever, pain, and a perceived mass. The objective time-space dimensions, in which the evolution of the disease took place, were reprocessed subjectively by the patient, producing an incorrect history. The objective time scale, whether for the history of the illness or for a projected future, *merely provides the language for time*; only with the patient can the *meaning* of the time arise, and the meaning is what the patient will use to organize actions or fears.

As in all dimensions of subjectivity the person's sense of time-space can change. Some changes in time sense, from childhood to old age, appear developmental; but events in one's life can also produce change: "What my illness taught me is the value of each day." Although it has not been studied, the time-space continuum of the very sick seems to me to be even more idiosyncratic than for most, and approximates the manner in which children deal with time (Piaget 1969). Thus, when listening to a patient's history off illness, we are always dealing with *two different* time scales: clock and calender time, and subjective time. Since the words are the same, and because we, within ourselves, are doing the same thing without being aware of it, it is difficult to keep the two kinds of time separate.

The Assignment of Cause to Experience

Just as no event can be experienced without a search for meaning, so too no event can be experienced without a search for cause. This is the dimension of causality, of cause and effect. Events are not only caused, they cause other things to happen: they have effects. Our conceptions of objects, events, relationships, and people, as well as the event that is a body sensation, include beliefs about where the event comes from and what will follow. When someone develops a pain that is attributed to carrying a heavy suitcase, more goes with that assignment of cause than merely an explanation of source. Cause suggests prognostic implications as well. Such pains usually come from "muscles." I put quotes around the word to distinguish the patient's word, "muscle," from the technical word, muscle. "Muscle" pains are usually short-lived, require no treatment, and herald nothing serious. (Unless the patient is a ballet dancer, a long-distance runner, or has dermatomyositis—all of which would change the prognostic implications.) As is the case with the space-time dimension and other facets of meaning, the understanding of the cause of one individual event does not occur apart from a person's whole pattern of causal understandings. To believe that one's pneumonia was caused by a bacterium or virus, one must conceive of microorganisms as existing and as being part of the general class of causes. If this knowledge exists, it may be applied by patients to a whole range of phenomena experienced in their bodies. Thus the alien body sensation may be subsumed under its cause, as in "I had a virus last week." Physicians also have a causal nexus, but our understandings may be different. To us, virus may mean, for example, coxsackie, varicella-zoster, reovirus, or mumps virus, in their most specific technical sense. Physicians also often share with patients the more general and vague sense of virus. For instance, "virus" may be synonymous with "an illness of unknown origin that has provoked a non-life-threatening immune response."

This distinction should allow me to make clear what I mean by cause as a part of a person's beliefs and meanings. The mumps virus, in the doctor's understanding, is a concrete object, having specific characteristics, including spatiotemporal dimensions. By that general class of causes known as "a virus," no such specific virus is meant. The cause, "a virus," is not concrete in itself but is functioning more like a symbol for "nonserious illness." For example, "viruses" always have antecendent causes, such as "I was run down," which have their own antecedents, as in "I was under a lot of strain." The train of anteced-ents may ultimately lead to some psychological conception of disease causality. In another person the ultimate set of causes may be self-blame and sin. The first words of one of my patients found to have carcinoma of the breast were, "I knew it, I'm being punished." Just as it is as useless to argue with people about their time sense or their values, it is equally useless simply to tell them that carcinoma of the breast is not punishment for some misbehavior. These aspects of belief and meaning are *part* of a person; to change them you must change the person.

"Viruses," in their common usage (as opposed to viruses, in their technical sense) are not considered serious, and so they are not usually seen as causing other serious events. (Linguists speak of the everyday beliefs attached to the word "virus" as being part of the connotative meaning of the word; the objective characteristics in time and space are the denotative meaning. The important thing for someone taking a history to realize is that when the patient says "I had a virus last week," very little is being said about the sense data of the experience, while a great deal is being said about that person's ideas about causality. The doctor's reply should almost always be, "What do you mean by virus?" To believe that the word "virus" has the same meaning to the patient as to the physician may lead to serious diagnostic errors.

When someone assigns a nonthreatening cause to an event (like "virus" or "muscle"), the happening is, so to speak, put to rest. By this, I mean the event can be put out of mind. Obviously, when serious cause has been assigned—"Do you think it's cancer?"—then the event *cannot* be put out of mind. Around and around fly the patient's thoughts, gathering and dismissing evidence relevant to the dread cause. As I said in volume 1, it is one of the foibles of the human condition that when a person could equally consider a dread or a benign cause, the "fatal premise" always wins. But in the process of weighing the evidence, some of the information from the sense organs will be dismissed because it does not fit the causal conclusion, while other sense data will be enlarged in importance. Thus when a symptom is reported to physicians, we get the result of the mental deliberations. It follows that the report may be solely, "I had a virus last week," or some other diagnostic category, as in the lady with the diverticulitis. Patients rarely report, however, that "I discovered cancer last week." Yet they may believe it to be the case and report only those data that agree with their covert conclusion, while disregarding contrary experiences.

It should now be clear why the history *as given* is almost always valueless and can in fact mislead. It could not be otherwise, even if Sir William Osler were the patient. *The patient does not have the option, nor the interest, to relate things objectively where illness is concerned.* In order to process the sense data of experience, meaning must be assigned. When reporting the details of timing and causality, and when choosing the words to describe events, the patient can only reflect the meaning of these events in his or her own spatiotemporal, causal, and value terms. A month may have thirty days, but which is longer, thirty days of health or thirty days of pain?

Interestingly, when a person does spontaneously report alien body sensations without assignment of personal meaning, the accuracy of the report about the body can be startling. I asked a woman with left upper quadrant abdominal pain what she thought the trouble was. (That is a useful question, because patients are frequently correct, and if not, it is a very direct way of finding out what worries them.) She said, "Well, of course, I know it's heart disease, because my whole family has heart disease. But sometimes, when I think to myself that maybe it isn't heart disease, then I think, well maybe my kidney is a big cyst and it's bumping against my ribs." Bull's-eye! That was precisely correct. This was even more surprising, because years earlier she had been told that the left kidney was atrophic. You will encounter descriptions like this more than once.

Patients tend to report events in one of three ways. The first is as a bunch of symptoms that, to physicians, may have nothing to do with one another. In the second mode of presentation, the patient gives the report in diagnostic terms. In the third type, the story is presented as a causal chain, where one event follows another, as though the first caused the second, caused the third, ad seriatum.

Here are some examples where the patient reports a bunch of symptoms, usually with a secret diagnosis in mind. The patient may not be sure what it all means, or perhaps he or she has several possible meanings in mind and cannot decide, or perhaps one *bad* possibility has raised its head, and the patient is bringing forth evidence to counter the *bad* meaning. One cannot know in the abstract, but when you take the history and start asking questions, you can find out.

Now then. Do you want to tell me the story again, please?
Well, briefly, I put off— I've been having a reaction, you know, in taking the Synthroid. Thyroid. And I began to have what seemed to be arthritic pains in the mid-region. And I finally realized that it could be the exer— that I was using dumbbells and exercising this part of the body because it's gradually getting better.

I started having the pains immediately after I started taking the thyr— I would get up— When I tried to get up, I'd feel stiff, and I've always been very flexible. And then I began to actually have pains and, you know, really sort of have to hippety-hop a little bit when I first get up. You know, over a period of time. (I need a few notes because I don't think I included some of these things in the medical history. And when I called you, I only told you the one thing that I had this shortness of breath.) But I had delayed all these things because I— knowing that I was doing more than I should be doing, you know, an unusual amount of work— physical exercise. I realized that it could be that, and I was waiting to see what did happen.

But the shortness of breath has seemed to be a little worse than it was— or happening more frequently, the shortness of breath. And the other thing which I think I should tell you about is a kind of wheezing sound that I had. This I first noticed after I had been doing a lot of heavy housework, and I went on vacation. The first night away, I woke up, and I had a— I could hear— I wasn't making a noise from my—
Doctor makes wheezing sound.
Yeah. But could hear this. Then I've only had that once since.
You mean the noise?
The noise. And then not as bad.

This patient was in her first episode of congestive heart failure with atrial fibrillation. The ache in the middle, the Synthroid, the barbells, the "hippity-hop," had nothing to do with it. If it had been her second or third episode, she would, in all probability, have given shortness of breath and palpitations immediately. The patient can be quickly trained by previous illness. In fact, if the patient has had long experience with a disease, he or she may be *more* knowledgeable about the symptoms than the physician. Sadly such patients often report that their doctors pay no attention to what they say about their symptoms. These physicians probably never owned an automobile. Every driver has had the annoying experience of telling an auto mechanic that the car made a *garee-peta*, *garee-peta*, sound when shifted into reverse, only to have the mechanic dismiss it as impossible. Then follows a large bill for (futile) repairs to a blameless water pump. The patient whose experiences have been brushed aside feel just as annoyed and frustrated as you did when that happened with your car. The troublesome thing for sick persons is that they cannot go back to the garage for another new part! The best teacher one can have about how a disease expresses itself in the person is an articulate patient who suffers from it. (An even better source of instruction is having the disease yourself. Unfortunately such a teacher may kill you.)

One hears the complaint that most patients have nothing but emotional problems: "You never see any interesting disease in a doctor's office." Not true, there are all sorts of interesting cases. But patients with fascinating diseases do not enter the office with a sign that reads "scleroderma." (Or, as my interns used to say, "fascinoma.") Patients usually present a whole array of symptoms, and buried in the middle is, "And my skin looks shiny on my hands, and it's hard to bend my fingers," followed by flatulence, itchy scalp, and athlete's foot.

This next patient is a fifteen-year-old girl, who also presents us with a list of symptoms.

Yes, Ma'am. What can I do for you?
Well— For about a week now, I've been—my stomach's been bloated, you know, really blown up. And I throw up, and I'm constipated at the same time. And so it's— it's really bothering me. I mean, I thought it would go away faster, but it hasn't gone away. And so—I just get these ATTACKS, of throwing up. And I got— I— I have been, you know, throwing up a lot lately. Just this year, since— well, August I had an attack, and I just would throw up and throw up. And it's not normal throwing up, though. Usually, you know, you throw up from DOWN. But this was kind of throwing up from— from right here.
Under your breastbone, there, mm-hm.
Yeah.

We will meet her again (and several of the other patients as well) in a subsequent chapter, as we see how questioning makes a case unfold.

This next case is entitled "Forever More a Mystery," because that is precisely what it turns out to be.

I went out—I was out at my beach house, and I went out on an all night drunk with my next door neighbor. And the next mor— I have no idea what happened, but the next morning I woke up with pains, particularly in my shoulder. I couldn't use the sh— shoulder muscle at all. Which got progressively worse.

I was bedridden for about two days. Muscles all over me started aching, started getting hot and cold flashes, a complete loss of appetite, fainting started developing, trembles and shakes; fast, shallow breathing that would go on sometimes for ten to fifteen minutes, and that went on close to a week. Meanwhile, after two or three days, a cough developed which was a big problem because my left side of my chest and lower back muscles were very sore and every time I coughed, that would be very painful.

Finally, they started subsiding, and I was in extreme lethargy throughout all of this. I got up enough energy— I had made it to the office, but by Thursday I was actually able to do some work which was an improvement over what I had done for the three prior days. A fairly unusual amount of lethargy remained Thursday, Friday, Saturday, Sunday—a good deal less active than I'm accustomed to being, not withstanding the fact that I was getting back to a fairly normal eating habit by then.

Muscle pains—I've been taking muscle relaxants of various types through all this—those names I don't know.

—do you really not know what happened that night?

I really don't.

Now, who'd you go with?

My next door neighbor, and he was in exactly as bad of shape.

And did you drive?

Oh no. No, this was Fire Island. It's boardwalks. I suspect—well, I suspect I didn't fall because there would have been bruise marks, I suppose. I suspect what happened is I passed out and uh, someone was trying to carry me and did a very poor job, and just—

Um, well, everything—just to finish my story—everthing seemed to be on the way back to being all right until Sunday night when I woke in the middle of the night with what for me seemed like very violent nausea and diarrhea. But I stayed home Monday, and I went to see my GP Monday night and got the Lomotil and other stuff. At the moment I'm not taking the cough syrup—the cough seems to have gone away. I'm just taking the Lomotil and the stuff that tastes like banana—Donnagel? And the Traxine—one of the three things I'm taking right now. I— so the only symptom I'm really getting is still what, for me, and I've eaten quite substantial meals in the past couple of days and getting twelve to sixteen hours of sleep a night, which, for someone who's accustomed to getting five or six is quite a bit. And I'm still finding myself quite tired. But other than that, the only problem I have is— is continuing pain on he left side of the chest whenever I try yawning or breathing deeply. My initial suspicion about a week ago was that I had cracked a rib. And, ah, the— I'm a little freaked out, which is how I got the Traxine because I have never before come across such an enormous assortment of symptoms—

Symptoms. Yeah. I can understand that!

And I do not have a history of hypochondria, so it's a little weird,

I find it difficult to focus my attention on this history. Where is the story line? The patient's list of events and symptoms are like papers tossed in a pile. One thing leads into the next, but the underlying premise is difficult to extract. Listening to such a history may be boring. You may find your attention wandering, again, and again. Repeated questioning does not seem to help matters. Before you blame yourself for not doing a good job, ask yourself why you are bored. Boredom is different from being tired or sleepy. Boredom is an active state. You may discover something about the patient that is actively boring. It may be the patient's way of throwing you off the track, or some other psychological mechanism coming into play. It may be none of these, but the feeling of boredom should alert you to look for its source. If nothing else, the search will relieve your boredom.

Sometimes histories of illness *are* very long and complicated; especially that of a hospitalized patient who has had multiple previous admissions. In these situations it is best to find other sources of information, such as old records or reports. In the office setting I frequently ask such patients to return the next time with the whole story written down, including names and dates. On the initial visit I try to find out what *today's* problem is and then concentrate on one manageable aspect of the illness, while awaiting the more complete record. This also takes some of the pressure off the patient, allowing the story to be recollected at leisure, and with the help of others. When it is unavoidable, as on a first hospital admission for a long, long illness, then one must patiently separate out each thread of the story and provide the patient with guidance for the telling of it. Becoming impatient merely flusters the patient and, in the long run, makes the whole thing take even longer.

The second kind of presentation alluded to earlier is the "diagnostic pronouncement." There are patients who tell you virtually nothing; they just announce their diagnoses. The worst thing about this is that you are liable to listen and believe!

Tell me the story.

Early in the spring, I had amoebas. OK.

Where'd you get 'em?

I don't know. I had had them before I went out of the country because I had all the symptoms. And then I went on vacation.

On the vacation I barely could move, and then when I came back, I decided to have it looked into. And it turned out to be fragilis. I lost a little weight; I was down to about 104 from about—110. I'm 106 now. Got over the amoebas. Had a particularly fatiguing time.

I am pleased that she "got over the amoebas," but I wonder what she had. My scepticism may sound as though I am now doing precisely what I criticized only a few pages back: dismissing what the patient says as unimportant. Not at all. I believe everything everybody says about anything. I only want to know exactly how they know. Who told them; what test; what did they feel; when; had it happened before; was it *exactly* the same; if not, why not; and on, and on, and on. I take it as a matter of principle that *nobody* knows better than the patient what the patient experienced. But even the most articulate patient may require assistance in exposing every dimension of his or her own experience.

This next example is similar.

I had an attack of diverticulitis in January. The first that I have ever had, although I have had a LOT of stomach trouble. Sort of all my adult life. And since then, I've been only fairly well. I remember very well the onset, because it was January 1st and I went to a New Year's Eve party, and I had this terrible feeling of something. Just real pain, right there—

High up like that?

Yeah, in a way. It didn't seem unlike as my usual. And that particular pain, I had had—I remembered one day having a meal at Green Mansions, I think it's called. Coming out of the restaurant and having that same pain. And I came out of there and I remember thinking, "Oh, goodness, I shouldn't have eaten that carrot salad." And that was last summer. Then I had this other one on January 1st. And then from January 1st on, I was really— I had lots of problems, this pain in the waist—and finally, I think it was January 22nd but I went to the doctor.

The visit with this patient took three hours! It took me so long to get at the real problem that I would talk to her for a while, have her wait in the waiting room while I saw someone else, then work with her again. But I finally found out what the matter was. I have no idea how long she was in the office. The typed transcript of her visit was eighty-three pages! For those of you who cannot tolerate falling far behind your schedule, the best thing is to ask the patient to return when you *do* have time. The patient will be grateful to have somebody who really listens and will usually accept the suggestion. In my office I keep shifting such patients around until finally everybody else is gone, and I have fewer time pressures. If you cannot hear the whole thing out, the chances are that you will miss the true problem and the patient will end up consuming even more time in subsequent wasted visits and unnecessary tests.

Often the most interesting histories to listen to are those that present a causal chain. They do not get you closer than the other type to finding out the patient's problem, but they provide an insight into the patient's causal beliefs. For those of us stuck in our offices, patients like the next one bring in a view of the wide world outside.

Tell me the story of your illness.

Ah, well—

You look lousy.

I do. I know I do.

Well, we got back last night very late. We were on the airplane such a long time. So anyway, I was fine in Peking. We were there a week. Then we go to Tientsin. Cold, very cold, it was. It was colder inside of the hotel than outside. All right? And I just couldn't take it. And I just started coughing and coughing, and my Charles put up a temperature, which I think I did not. And I stayed in bed! Three days! Coughing. Coughing. And no strength. OK. Then, big, heavy, thick, ugly things, right? Listen.

Then we took the train to go to Shanghai. That was almost like a twenty-three-hour train ride. And we did— yes, I was not feeling well. I was still cold, no heat in the rooms. They don't put the heat on before December 15th, right? And so I took myself into the hospital. And they told me I had bronchitis. And then so I stayed in bed for two days. And then we got to Canton, southern China, where it's supposed to be hot, and it's not hot there because they're preparing for a typhoon. So it was very cold, windy, and I had to also end up going to a hospital there because I was coughing. I couldn't stop coughing! I would cough for a half an hour! It never stopped. And I was not getting phlegms any more. It was all up in my head, and in my throat. It was all cleared in here. So I was feeling better, but that evening—

(Doctor coughs.) See, I got it already!

(Laughs)

Anyway, they gave me antibiotics then. So November 16 we got out of China. Then I had a lot of fun in Hong Kong, where it was hot finally. No, but it was not fun, either, because we would go to a restaurant, and they would have two strong air conditioners. And then I started getting the chills again. And then for three days in a row I'm having headaches. And my ears are feeling funny. You know, like they're always running.

And so I went to see the doctor, and he said I still had bronchitis. And he gave me a shot of antibiotic, and I was supposed to have two. Two days in a row. And I passed out from the shot. So he got panicky.

You what?

I just fainted. When he gave me the shot, the nurse gave me the shot, I felt fine. I got up, went— in— in the rear end she gave me the shot. But at her desk, waiting five minutes, standing up, for her to give me some more prescription and then walked out of there, moved to the elevator, and I felt it. I was seeing everything black, and my ears, I was hearing funny noises, so I ran back into the doctor's office, and I just (gasp) went off! And I know that he carried me to the couch like this, and he covered me right away and told me to keep quiet for fifteen minutes at least. And gave me some pills. But he did ask me to return the next day. For another shot. But I think in the meantime he though about it because I got there and the nurse said, "No, no. No more shot." And that's my story. (Sigh)

Here is another of the same variety.

How can I help you, Ruth?
Well, gee. You see— see, I was performing in Munich, and it was raining all the time. I mean, I never got any sleep. It was like I was running around with Susan and Sally and Bill, who were all sick, and the weather was damp and disgusting and raining all the time. And I was— I was under a lot of continual pressure, and I had to keep shlepping all this goddam theater equipment from place to place. I mean, every time I went to the train it was pouring like forty days and forty nights, and I was exhaused. And it was like that from Munich to Dusseldorf, where Whitney got sick. And of course I was in the same room with her, and we were running around like refugees all the time. I'm surprised I don't have the plague! Of course I don't actually *have* the plague. I don't think.

And another.

Sometimes it might be that I'm kind of tense at work, and I'm swallowing a lot of air and I sort of gulped down a yogurt. But all through this, if I've had— I usually have a yogurt for lunch and half an apple a little later, sometimes I'll have gas pains during the afternoon lasting—not very bad, pretty mild, lasting maybe an hour or two. It's fairly frequent.

Once again, little is presented that would stand on its own, in diagnostic thinking.

If the central point of this chapter is that the patient's history as given is of little diagnostic value, why is it received wisdom that the patient's story of illness is so important? First, no one except the patient knows what happened, so no matter what is said or how it is told, the history contains information otherwise unavailable. Second, as noted early in this chapter, the patient's meanings and the doctor's meanings are often so close that, without being aware of it, the doctor fills in the gaps in the history and acts as though the patient had furnished the information. This next example, of a garden variety cold, should make that point.

How can I help you?
Well, I got sick on Christmas, and I'm not quite over it yet. It started as a sore throat, and then it got into a cold and cough. And when we were down South and when we got back, it developed even more, and I ended up with severe sinus headaches, which I have had before. But usually I've had somebody to tell me what to do about it. So Joe had had the problem before, and he told me to get a vaporizer to use in the room. So we used the vaporizer in the room for almost two weeks, and I haven't had one of those headaches for almost a week now.

Even though physicians and patients share many meanings, "close only counts for hand grenades." Close is not close enough for history taking. Thus, to get at the real story of patients' illnesses, it is not sufficient to peel the experiences away from the patients' assignment of meanings, we must be doubly careful to keep our own meanings out of the story. In its crudest form the leading question interpolates the physician's meanings. But much more subtle contamination of the patient's story occurs when the physician believes he or she knows what the patient is truly saying, when actually the facts remain obscure. The dictum, "Let patients tell their stories in their own words," is to correct the propensity of physicians to tell patients what patients are saying, instead of listening.

The third reason that histories as given frequently seem adequate is that patients are often trained to tell their stories in a manner acceptable to physicians. Listen to this next patient with a cold and contrast it with the previous example.

How can I help you?
Well, I had a pretty bad cold. I think it was going around with everybody at school. I think it started about three weeks ago.

I remember waking up one Saturday morning with a terrible headache like a head cold. And the symptoms were more like what I would consider a cold—the runny nose and the achiness has gone away. And I still have a cough, and it feels like it's deeper in my chest. And what concerned me the most was also I had a lot of pain in my ears and a headache, I guess, above my eyes.

So my father suggested that I come to the internist first, have it generally checked out, and then see what you recommend from there.

See how much to the point her story is. It ought to be—she is the daughter of a physician: a trained patient.

The final reason the wrong belief prevails about the patient's history is the most common reason for ignorance—the subject has not been studied systematically.

You have had the opportunity in this chapter to read twenty-two histories of illness in their natural state. I have presented an account of why patients behave as they do. This is the way we all behave; all of us assign meaning to events, objects, people, and relationships. The patient's history achieves its central role in medical care, not in the form as given but as a result of the interaction between physician and patient. The art of history taking, which will occupy the next three chapters, is the art of questioning, and then listening to the answers, in order to find out exactly what happened and to whom. Our questions will be: What is the disease? And, who is the sick person? Until both answers are known, the doctor's job of making sick people better will not be properly started.

References

Cassell, E. J. The subjective in clinical judgment: a critical appraisal. In *Clinical Judgment*, H. T. Engelhardt, Jr., S. F. Spicker, and B. Towers, eds. Dordrecht, Holland: Reidel, pp. 199–215.

Percy, W. *The Message in the Bottle.* New York: Farrar, Straus and Giroux, 1975.

Piaget, J. *The Child's Conception of Time.* London: Routledge and Kegan Paul, 1969.

Zborowski, M. *People in Pain.* San Francisco: Jossey-Bass, 1969.

2

Asking Questions about the Body

History taking is often taught as if the object is to strip away all the confusion heaped on the facts by patients in order to get at the diagnosis. The trouble is that when doctors discard all the patient's meanings, values, causal notions, and time-space distortions which cloud the particulars, they may have found the disease, but they have discarded the patient! On the other hand, physicians concerned with patients as persons who exclude a relentless pursuit of the specifics of the disorder may know the patient but not the diagnosis.

The goal is to find out *both* what is happening in the body—the pathophysiology of the illness—and who the patient is. Then the two kinds of information must be integrated so that the doctor can know how this *specific* illness came about and how this *particular* sick person can best be helped. The process is initiated through the personal interaction between doctor and patient known as history taking, which employs the special instrument known as questions. This chapter is about asking questions.

Some general comments are in order about this unique interaction between doctor and patient. Taking a history is, above all, a *cooperative act*. There can be no accurate history without the help of the patient. Patients come to doctors because they want to be better; therefore they have a natural desire to help you. It is absolutely essential that you never forget this because patients—all patients whether strangers, friends, doctors, enemies, or allies—may resist, obstruct, lie, deceive, conceal, annoy, antagonize, fight, flatter, forget, be seductive, play hard to get, be hostile, be overly helpful, be absolutely silent, never stop talking, some of the above, all of the above, occasionally, sometimes, often, forever, or never! Why do people behave like this? Because they are patients and because they are sick. Most people are lucky that they have never been seriously ill, and thus they do not know one of the basic truths of medicine: sick persons are different from well persons. One of the worst things that happens to sick people is that they are no longer in control of their world. The loss of control creates such dreadful stress that people will do almost anything to remain in command. Patients who can no longer manage themselves may attempt to remain in control by manipulating their doctors. One way of accomplishing this is by refusing to give a history, or by "rewriting" history in the telling of it. If you always remember, however, that the patient really wants to be helped, such tactics need not throw you off the track.

I am aware of how tough it can be to stick to the principle, and how long it takes to learn how not to struggle or fight with patients. But, as they say in the treatment of that most stubborn of diseases, alcoholism, partial success is not failure. It is useful to remember that when patients argue, object, yell, scream, fight, or anything else from the rich repetoire of behavior that can turn caregivers into combatants, they must have a very good reason for doing it. It is possible to use their resistance to help overcome the obstruction and move ahead with your history. The obstructionism can often be surmounted merely by acknowledging its right to exist. As you walk to the patient's bedside, the dialogue often goes like this: "Mr. Smith, I'm Dr. Jones, and I'd like like to find out the story of your illness." "*What, another doctor—you're probably not even a doctor. I've had enough. Go experiment on somebody else!*" Oh dear, this is a difficult opening to overcome. Now, on the defensive, you might say, "Mr. Smith this is very important and, after all, it's for your benefit. Besides, it won't take long." Mr. Smith has all he needs now: "*Don't you think I know what's for my good or not? Just 'cause you're a doctor doesn't mean you know everything thats right for me. Besides, the last doctor who said it wouldn't take long was here two hours.*" Typically Mr Smith will finally give in, but the history-taking interaction will be tainted by the interchange and the anger. In addition you will probably be afraid of really going after the information you need, for fear of arousing Mr. Smith's anger. There is, however, another way. You might have said, "Boy, Mr. Smith, you certainly must have a good reason for being so mad. Nobody would be as upset as this without a mighty good reason. Nobody could blame you for being as worked up as that, with the reasons you must have. Certainly not me. Why don't you and I do as complete a job as possible to at least make all this trouble worthwhile." Believe it or not, Mr. Smith will probably become a most cooperative patient, especially because you have been so understanding compared to most of your colleagues. Please notice I did not use the word "anger." I employed "mad," and "upset," thus avoiding the retort, "I am *not angry!"* Also actively avoided was inquiring why he was upset. His reasons are not unimportant, but more important at this time is acknowledging his right to be upset and then getting on with the history. If he insisted on explaining, that would be different. But, if he does, *do not defend* the kitchen for cold meals, the nurse for delays, or your colleagues for errors. Simply reiterate, "I don't know the facts Mr. Smith, but I sure know you wouldn't be so worked up without good reason." Then start taking a history.

For those readers new to history taking, a few simple precautions will make the process go more smoothly. Make the patient and yourself comfortable. Be sure that if the history takes a long time, the patient does not become fatigued because of his or her position in the bed or chair, or because your back is to the window, making it necessary for the patient to continue to look into the light. (Placing a little sign on your clipboard that says "Is the patient comfortable" will help you remember, and remind you to stop from time to time to arrange pillows, pour the patient some water, or do some other act that will increase both the patient's comfort and cooperativeness.) Sit facing the patient so that you will have no difficulty maintaining eye contact. Nothing denotes interest so much as looking at the speaker intently while you listen. Take notes, but do not bury yourself in them. If you forget what you want to ask, do not hesitate to pause and look at your notes. If you think this indecisiveness indicates ineptness, you are correct. In fact you are not skillful; how could you be? But the pause will help you collect your thoughts and ask the proper question. That way, you will take a better history and come to be proficient much faster than if you merely tried to preserve the appearance of competence.

What Is the Story?

The time has come to pursue what is happening to the sick person's body. All diseases, all illnesses, play themselves out over time. They are an unfolding story of changes in function. As the story evolves, the task is to understand what is happening in the body and to form an hypothesis. It is very important to resist the pressure to name the disease. If one begins to believe in a diagnosis too soon, then contrary evidence may wrongly be put aside. The hypothesis initially should be of a general nature, such as "This is musculoskeletal," or "I think the problem is cardiac," or "It sounds like infectious disease." These may not seem like very sophisticated hypotheses, but they provide a place to start. Gradually the hypothesis should become more specific.

Once again, I believe one does better with the hypothesis "inflammatory bowel disease," than with "ulcerative colitis"; similarly, better with "dyspepsia," "indigestion," or "acid diathesis" (call it what you will), than "duodenal ulcer." The reason is that the hypothesis is meant to guide the diagnostic and therapeutic actions necessary to making the patient better. If one believes that the patient has a duodenal ulcer, a GI Series may follow. But the history of gall bladder disease is sometimes indistinguishable from the history of duodenal ulcer, so that gall bladder disease must be included in the diagnostic thinking. If you think in very narrow terms of esophageal acid reflux ("hiatus hernia"), how long will you wait before you decide that the patient's failure to improve is not lack of compliance but the wrong diagnosis? It should be obvious that I am trying to move your diagnostic thinking away from the usual focus on diseases and toward a primary concern with the events in the body that lead to the patient's present state.

As I suggested before, the unfortunate but very important thing about hypotheses is that they are very difficult to prove correct. It is far simpler to prove the negative or null hypothesis. Therefore, in taking a history, *you should always attempt to prove that your hypothesis is wrong.* Every answer that seems to support your belief should be counterquestioned to see if it can be undermined. You are not after the *truth*, you are after a belief that after repeated attempts remains unassailable, and it will provide a *basis for the correct action.*

One further note, before we examine the process of questioning. This chapter is concerned with the single-minded pursuit of what is wrong in the patient's *body*. At times you may be surprised by the forcefulness of the questioning. I do not wish to soften that impression, because obtaining the information about what happened to the body may require the persistence of a bloodhound. It will not be until the next chapter that the same single-mindedness about who the patient is will emerge. In practice, however, the two are always occurring concurrently so that the diagnostic act of taking a history is, at the same time, therapeutic, showing the physician as someone who cares about *this* sick person and how *this* person is to be helped to become better.

Often the hypothesis can be formed from the patient's initial statement. Thus the first question should elicit the most useful opening remarks. In the office I say, "How can I help you?" or "Please tell me the story of your illness," or "Please tell me what the problem is." In the hospital, "Please tell me the story of your illness" is a good opening question. When the problem seems complicated, or of long duration, it may be helpful to ask, "When did you last feel entirely well?"

Having been asked to tell the story of their illness, most patients will do just that. A rather long opening statement usually follows, during which the physician should actively listen and watch. This provides an opportunity to hear, not only the content but the word choice, syntax, logic, and paralanguage, and to observe the patient's face and motions. The opening statement may also be short and to the point, as in this next example. This is only the first segment of this patient's history; more comes later.

How can I help you?
OK (Laugh) Um, I've had bleeding
Mm-hm.
right before my bowel movement.
Mm-hm.
Some diarrhea and then bleeding.
Mm-hm.
For about—it was on and off and it went away. And then it's been consistent for the last eight days.
Mm-hm. And when is the first time that ever happened?
I really don't remember because I didn't make any note of it.
OK.
It was just like a little diarrhea and no blood—I think I had hemorrhoids.
Mm-hm. When was that?
But, um—about two months ago.
Mm-hm.
But and, it was—um, but then they went away. I didn't do anything for them.
Did you have bleeding back then, a few months ago?
No.
I see. And how did you know it that time that you had hemorrhoids?
I thought they were because I felt something there.
Did you feel something? A lump or somethin' like that?
Yeah, Like two little lumps.
Right. Now. Did the diarrhea come first?
Yeah.
And when was that? Roughly speaking.
About—a month ago, maybe—

Mm-hm.

But then it was on and off I— and I did—

You mean for one day you would have dia— a loose—What do you mean by diarrhea?

Not diarrhea. I don't even know! Like, I can't describe it—

Mm-hm.

Like, I would feel very gassy.

Right.

And, like, I guess expel— like, I don't know if it was mucus or whatever.

Mm-hm. You mean you would go to sit down and just mucus or gas would come out?

Yeah.

But you had the feeling before you did that that you were going to have to move your bowels?

Right.

And then would that happen again in a couple hours or a few minutes or something like that?

About once a day—

Mm-hm.

(whisper)—or so. Yeah.

What can the matter be with this woman? Bowel movement abnormalities are a good place to start our exploration. Because all of us have bowel movements, we all have a framework of reference to help us understand her. Persons taking histories should use themselves and their own experiences with their bodies and the world as a reference for what they hear. This has an obvious limitation, especially for the young and inexperienced. What happens if you have never had the experience or felt the body sensation? You simply have to ask more questions; enough so that when you are finished you have both acquired the diagnostic information and learned more about the world. It is vital to emphasize what I do *not* mean. *You are not using yourself to judge whether or not the patient is telling the truth.* You are checking with your own knowledge to be sure that you understand the symptom completely; understand it so well that you can, so to speak, feel it within you. This concept is central to the entire process of obtaining a history, so let me phrase it yet another way. You are taking the patient's articulation of an event and translating it into your language. The repeated questioning allows you to check your translation. You use your self-knowledge and experience to check if you have clearly understood what the patient means, and you use your knowledge of anatomy and physiology to help guide the questions. For the example just cited, knowledge of the anatomy of the rectum and dynamics of defecation are necessary.

What is happening to this patient that is different from a normal bowel movement? She has had some bleeding before her bowel movement and then some diarrhea. She thought the bleeding was hemorrhoidal, because two months earlier she had hemorrhoids. The doctor asks how she knew this. He is aware that people very commonly feel their own rectums, especially when bleeding occurs. He also knows that hemorrhoids may feel like soft, raisinlike lumps, so he uses a leading question, "... a lump, or something like that?" If he did not have such knowledge before treating this patient, he would have begun to acquire it from her.

Soon after the patient reports diarrhea, the doctor checks to find out what she *means* by diarrhea. The more common the word in question, the more important it is that you check what the patient means by that word. Indeed, she only used the word because she did not know what to call what she experienced. She felt gassy and then sat down and expelled some mucus. Most people do not sit on a toilet to expel gas, but almost everyone has had the experience of going to the toilet because of not being sure whether the sensation of the urge to stool heralds flatus or stool, and one wishes to take no chances. The mucus probably originates very low down in the GI tract. Remember, she does not have mucus in addition to a bowel movement, or a bowel movement and then mucus. She experiences an urge to defecate, and only mucus comes out. Since the sensation of a need to defecate arises from pressure in the rectum, the mucus must originate in the rectum or not far above it, since otherwise it would be expelled along with the feces.

Let us learn some more of her history.

And did you move your bowels normally AS WELL?
Then?
Mm.
I think so.
So, if I have it correctly, something new that wasn't present before is that you used to have the feeling that you had to move your bowels—
Mm-hm.
—you'd sit down, and then what would happen?

And then I'd have, like, I guess, diarrhea/mucus—I don't know what to call it; just a little would come out.

Just a little.

Uh-huh.

And then would it also—

I don't know if I really felt I had to meel— feel— ah, move my bowels. Sometimes I'd just feel ga— you know, that I had, you know, that I had to expel gas. And I'd go to the bathroom and the mucus or diarrhea would come out.

I see. And you say that would happen about once a day?

No, not— not regularly. It has been—

No, but I mean, when it happened, would it be just once in the day?

Yeah. Yeah, yeah.

Right.

Just like whenever anyone would feel some gas after they's had some—

Right.

—something to eat or whatever.

Mm-hm. Right.

But this wasn't just normal gas, it was—

Mm-hm. And that was a month ago. And then when did it become more regular?

About eight days ago.

Right. And eight days ago— tell me what the change was.

The um, diarrhea/mucus or whatever, would come right before, like, a few drops of blood and then a regular bowel movement.

More evidence in favor of the mucus coming from low down in the bowel: first mucus, then blood, and *then* stool. We can guess that the mucus comes from the ampulla of the rectum or lower, and a similar origin is suggested for the blood. But we need more evidence.

I see, I see. Let me go back to the fir— In the beginning, when the diarrhea/mucus, whatever; would it also be followed by a regular bowel movement?

Not then.

Not then. Then when would you have your bowel movement? Sometime later that day?

Right.

I see. OK I don't mean to be so snoopy—

No. I know. It's OK. (Nervous laugh) I never made a note of— a mental note of it until, it was just a regular—

Mm-hm.

—until I had bleeding, actually. And that was—
Right.
I just attributed the diarrhea to, just something I ate or whatever.
Mm-hm. And had it happened in the past?
Never. Never.
And most recently, when you would have this blood—
Mm-hm.
—would it stain the toilet bowl red?
Um—it was— I— there was blood and—
And would it be as much as a teaspoonful, or a few drops, or—
I guess just a few drops, not really that much.
Mm-hm. Right. All right. So you move your— you pass some gas and mucus,
Mm-hm.
—a few drops of blood,
Mm-hm.
—then a normal bowel movement.
Mm-hm.
Right?
Mm-hm.

Please note that the questions use her own wording for the symptom. The question has dissociated the mucus from the bowel movement. The suspicion that this problem comes from the rectum appears more justified. Where does the blood come from? Obviously not high in the GI tract or it would be tarry, nor high in the large bowel or it would be darkish, perhaps jellylike in consistency. And not too much above the rectum or it would be mixed in with the bowel movement. Perhaps it comes from the anus; but if so, why does it follow the mucus? Perhaps the motion of the anus following the expulsion of the mucus brought on the bleeding. (After all, she told us that she once felt two lumps.) Let us see what the patient can tell us.

Would there be any blood after that, too?
No. None.
All right. And would there be blood on the toilet paper?
Um—not really. A little. Just—
Some, or none.
Maybe just a little bit.

Right.

But nothing extravagant.

Not very— but I mean, the important thing was that there would be blood in the bowl. Right?

Right.

OK. Then you turned around that day—

Ah-huh,

—eight days ago and looked in the bowl—

Right.

—and by God, there was blood in it.
And that had never happened before.

Right.

The blood does not follow a regular bowel movement, and the blood does not seem to show up on toilet paper, although it stains the toilet water. (Her initial response suggests that it does not appear on the toilet paper: "Maybe a little" is what people say when they try to please their doctor. But if it had been a crucial point, more exacting questions could have been asked.) It is possible to say with greater confidence now that the blood arises from the rectum, or nearby, just like the mucus. What process might produce these symptoms? It seems fair to say that the mucus membrane of this patient's rectum is irritated. Were it the anus, she would have pain. Were the irritation higher, it would bring on more frequent bowel movements, or at least a bowel movement shortly after the mucus and blood. Further, the process may be progressing, since in the beginning there was mucus only, but in the last eight days there has been mucus and blood, and the whole business has occurred with more regularity. A crude hypothesis can be formulated: something is irritating the rectum, and it is spontaneously getting worse. A more specific hypothesis seems reasonable: inflammatory disease of the rectum.

You may believe too much work was required just to come up with such a general formulation. Remember, however, that the process which occupied so much print took only four minutes and then seconds to complete! Much more important is what follows this hypothesis. What should be done next with this young woman? No one would question that she should promptly be endoscoped, in order to visualize her lower bowel. That may seem obvious to you because she had rectal bleeding. But in this era many patients do not present such symptoms initially in the office. Rather, they speak to the doctor on the telephone. Consequently the decision about when she should come to the office and what should be done (and thus what, if any, preparation will be required) will be based on *the same process of history taking done by telephone.* Many physicians are too busy to see promptly every patient with ordinary diarrhea or every patient who attributes bleeding to hemorrhoids—especially patients who have a previous history of hemorrhoids.

The patient had inflammatory bowel disease which initially presented as granulomatous proctitis. The actual diagnosis followed the diagnostic maneuvers initiated by the history as given here. The illness provided a good example of how the process of taking a history clearly brought out the pathophysiology of the rectum. And equally important, we see in this case how formulating a pathophysiologic hypothesis is *separate from making the diagnosis of a disease.*

Notice that the types of questions are varied. They include yes-no questions (such as "You didn't do anything about them two months ago?"), content questions (such as "When is the first time that ever happened?"), open-ended questions (such as "You could feel something, a lump or something like that?"). We will see many better examples of open-ended questions and discuss their function later in the chapter. Learning to phrase questions and choose the most effective wording is the prime skill of the history taker. It is, however, not the only skill. Questions must have a goal or all the virtuosity in the world will be for naught. The goal of the questioning is to establish, with the greatest possible precision, exactly what is happening in the patient's body—what is the chain of events of the dysfunction—or, to repeat the word I have been applying, the pathophysiology of the illness.

Because the type of question used should be determined by the information being sought, it is fair to say that there are no intrinsically "bad" types of questions. There are, however, inappropriate, badly timed, or badly worded questions. I say this because physicians are often advised not to ask leading questions, in order to prevent the patient from saying what the doctor wishes to hear. As we shall see, leading questions are often essential and can be very productive when used properly. Listen to your own questions, and choose the words very carefully. If the wording does not sound precisely right, stop and rephrase the question. One wants to avoid questions whose form logically precludes a useful answer. "So now you don't have any symptoms?" Yes means no, and no means no, and both mean yes—

the patient is symptom free. When the same content is expressed as "So, do you have symptoms now?" then yes and no are distinct: no means no symptoms, and yes means the patient has symptoms. Another very frequent error is one of condensation: "Have you had measles, mumps, or chickenpox?" If the patient answers yes, which disease has the patient had? If no is the answer, how can one be sure the patient merely did not have chickenpox?

Similarly, one wants to avoid the "When did you stop beating your grandmother" class of questions. In medicine, an example might be, "When did you finally start taking proper care of yourself?" A question should almost never force a patient into self-incrimination. More important, one wants, at all costs, to avoid encouraging the patient to lie, as in "I'm sure you are taking your medication regularly, aren't you?" Many of us tell small lies in order not to look the fool, but once the lie is said, the patient must admit to lying in order to give the information requested. Above all, *the patient is not your adversary*, no matter how much you must struggle to get a set of facts. If the patient is having trouble with a question, rephrase it, break it into two or more parts, ask its converse, in order to facilitate an answer.

The next example gives us the opportunity to consider open-ended questions in more detail. "How can I help you?" is the most general such question. "Tell me how it started?" "And what happened then?" All of these are ways of facilitating the conversation. Even brief utterances such as "And then?" or "Yes, and?" may be sufficient to promote the flow of information. Then, see how the questions become increasingly specific. This patient is a thirty-four year old man.

How can I help you?

Ah, I have a chronic cold, and um, cough. And the last night it was real bad, and I couldn't sleep. And during the course of the night I was— I spit up blood.

Mm-hm.

I've had a lot of trouble breathing, and I've been wheezing when I breathe for the last two or three weeks.

How did it all start?

Um, the first time I noticed problems with my bronchial tubes, if that's what it is, was this time last year when I had the flu.

And what happened then?

I had a lot of difficulty breathing, and it went away eventually, actually, after a week or two. And whenever I did get a cold after that, I would have a symptom of bronchial congestion which I've never had before.

The patient's initial statement includes his diagnosis of a "chronic cold," and describes his hemoptysis, difficulty breathing, and wheezing. A great deal of information is presented in less than thirty seconds. But to give it meaning for physicians, much more is necessary. "How did it all start?" moves the onset back one year. A mention of the flu, followed by another open-ended question, which leads to the report that he wheezed a year earlier. Thus, within moments of the opening statement, we have reason to believe that there is more to this story, and it is worth pursuing.

And what does that feel like?
Um, hmm, it just feels like—there's just tightness in my chest, I guess.
Want to show me where— in the middle, or—
Right in here.
And do you wheeze when you have that?
Yes.
Mm-hm.
Also, I have ah—apparently I have allergies, and I sneeze a lot. Sometimes it goes away, and sometimes it turns into a cold.
And does that happen at the same time as the ah, bronchial congestion?
Yeah.
And it never happened before last year? Not even as a child?
Um, I had the allergy, but no, I never had— I might have, but not, not to my memory.
Right. And then you were all right again, and— no, did— you told me, and then every time you get a cold it happens?
Right. Since last year.
And how long will it last with a cold?
Well, the um, colds normally last five or six days. But, ah—
Mm-hm.

What is "the symptom of bronchial congestion"? Even if we have all had the feeling of "bronchial congestion," we do not know what it feels like to him. In the remainder of the interchange, see how the questions are directed at finding out exactly what the patient experiences.

But I guess—about that time or a little longer.
And has the wheezing or the tightness—
Well, even if I don't have a cold, I have wheezing sometimes. And I have postnasal drip and—
So, about how many times would you say between the flu last year and whatever this is this time, have you had that— periods of wheezing?
Um, twenty or thirty.
And what have you done for that?
Um, nothing.

The illness, as finally revealed, is quite different than one might expect from his opening statement. Twenty or thirty episodes of frank wheezing preceded by sneezing does not sound the same as an episode or two of wheezing occurring with respiratory infection. What finally "brought him to the office" was probably hemoptysis! What is the evolution of his disease? Allergies arrived in childhood and then subsided. (These details were not developed in this fragment.) He then experienced frequent episodic bronchospasm in young adulthood, in a word, asthma. It is very interesting to find out from someone with recent-onset asthma (reversible acute obstructive airway disease) precisely what it feels like to have bronchospasm: where the sensations are, whether body position or motion of the head or neck sets it off, its relationship (or lack thereof) to stress, various stages in the development of an asthma episode such as the very first sensation the patient notices. Questioning in this way can really teach one about this illness. Do not be constrained about prying; the patients will generally not mind and may even enjoy relating their symptoms in visceral detail. Do not, in other words, merely ask questions until you have filled out the questionnaire in your memory labeled "asthma."

In the last example there were many questions whose purpose was clarification. Often it is best to use the patients own words in this process, as when the physician refers to the wheezing, or the tightness, or whatever it is?" This is especially important when you have difficulty pinning a patient down to a clear description. In this next example I was unable to get a satisfactory picture of the severity of the patient's chest symptoms despite repeated questions. I insisted on using the word "wheezing," because I was unaware at that time that cough is a frequent (and sometimes, isolated) symptom of bronchospasm. Not until I incorporated the patient's own language into the question did the story unfold. It is a frequent error in history taking for the physician to use his or her word for a symptom, rather than the patient's. Insisting on equating angina with chest pain is one such example; since angina may be experienced as discomfort, squeezing, pressure, choking sensation, shortness of breath, or even tingling in the chest or arm, asking whether the patient has chest pain may be like fishing for trout with a bass plug!

Perhaps doctors insist on using their words because they are looking for *important* symptoms. They are not after any old chest discomfort; they are on the hunt for *angina*. Anything less will be discarded. Though it is crucial *not* to overlook serious disease, the big-name disease safari utilizes only part of the modern physician's knowledge. As doctors we know more than merely diseases; we know how the body works in health and disease. Students complain that their first-year medical school studies are not relevant to medical care, but they are wrong. It is in taking a history that the knowledge of human biology first comes to bear on a patient's problem. Though patients are also concerned about heart disease, they want to know how their particular symptoms relate to the disorder they have. To answer their questions, it is necessary to show them how the malfunction came about. Let me illustrate with shortness of breath, because it is so common and can be so important. In the following simplified description of symptom pathophysiology, notice the opportunities that are presented for specific questions that will not only help distinguish between the various causes of dyspnea but will also allow an estimation of their severity. I have employed generally nontechnical language, in order to demonstrate how easily pathophysiology can be explained to patients. The sensation of shortness of breath is experienced as an inability to get enough air and is usually expressed as "I can't catch my breath," "I can't get enough air," or "I'm short of breath." We know that this sensation can arise from very different mechanisms. In congestive heart failure, fluid backed up from the heart stays in the lungs and reduces ventilatory volume. With exercise, oxygen demand increases and so does the degree of failure. With some rest, the pulmonary edema is reduced, and the dyspnea subsides, only to reappear with further activity. Thus questions about activity can provide an accurate gauge of the degree of left heart failure and its progression. With chronic obstructive pulmonary disease, effective breathing capacity is reduced. Leaving bronchospasm aside for the moment, there is not sufficient lung capacity for the extra amount of air required to meet the demands of vigorous exercise. But if the patient maintains a pace appropriate to the oxygen supply, some exertion can continue. There is, in other words, a certain constancy in the relationship between activity and breathlessness.

When bronchospasm is prominent, either in asthma or chronic obstructive pulmonary disease, the difficulty in breathing may wax and wane independent of the effect of exercise. Asthmatics, whose chests cannot be sufficiently emptied because of the bronchospasm, still believe that they cannot get enough air *in*. Thus they gasp for more breath, which only worsens their distress since it contributes to air trapping in the chest. When they stop gasping (and also relax their shoulders), they feel less breathless.

When questions elicit information that seems discrepant, there must be a reason, which further queries may reveal. For example, people with emphysematous chests, like those with congestive heart failure, may have to stop while walking, no matter how slowly they go. Coincident with the perception of dyspnea often comes the belief on the part of the patient that he or she will not make it to the next destination. In this situation instead of slowing down, patients frequently increase their walking speed, in order to arrive at, say, the bus stop before they run out of air. This gives the impression that the dyspnea is progressive, despite what may be reported as a steady pace. Many drivers running out of fuel act similarly, speeding up in order to get to the gas station before the tank is dry.

When data from the history seem to conflict with data from the laboratory, the history is too often brushed aside. For example, a patient with chronic obstructive pulmonary disease and a one-hundred-pack year history of cigarette smoking was restricted by her breathlessness to virtual invalidism by the time she was admitted to a hospital. Her $pO2$ was 55, but surprisingly, her $pCO2$ was only 33. Repeat values were the same. The physicians considered her to have exaggerated her disability. It turned out, however, that forty years earlier she had worked in a quartz factory. While there, in common with several other workers, she became ill with what had been called "acute silicosis pneumonia" and was sick for weeks. Pulmonary function studies confirmed the predictable impaired diffusing capacity. When information from the patient is to be disregarded, one should be as cautious as in dismissing any other findings.

Many, perhaps most, patients who complain of being short of breath have none of these diseases. We may hear them sigh as they are telling their story, or they may answer affirmatively to the question "Do you have the feeling that if you could only take a deeper breath you would feel better?" After more questions, a physical examination, and perhaps some other studies, we feel confidant in saying that they do not have serious disease. The symptom is emotional in origin. The patient is happy not to be dying but wonders how emotions create shortness of breath. An explanation will go further than reassurance in solving the problem. Shortness of breath in this situation is often created in the following manner. A breath is taken, but before it is fully expelled, the person breathes in again. By the third breath there is very little room left in the lungs, and the person experiences the sensation of breathlessness. Then follows a long, sighing expiration, and the process starts again. Try it yourself. The patient will be grateful to be told how to make the distress go away. Such patients do not want only a diagnosis, they want to breath again. As time goes on, one becomes expert, not only in knowing how diseases express themselves in a history but how even the most trivial symptoms come about physiologically.

I cough!

You cough. And do you raise phlegm with this cough?

No . R-recently I have begun to raise a little phlegm occa— in the morning. Occasionally.

Is that something new for you?

It's progressively more so. It isn't really—yuh. I don't do it all the time.

Mm-hm.

It isn't a large amount of—

When you— ah, does your wheezing get in the way of your activity, exercise,

No.

or walking up stairs, or walking down—

No. No.

In this questionnaire, there are questions about getting shortness of breath when you walk.

Yeah. I— I've— I ha— right.

Mm-hm. And—

It doesn't really. I mean, I— at the times that I feel—

Would you run up the subway steps?

I nev— No.

Could you?

Yeah! Sure I could.

Would it stop you, your wheezing?

No. I don't—Look, I wheeze only from time to time.

Yes, but it— Since you came back in September, and certainly in the last week or so, you've wheezed enough so that your wife is gettin' up at night, or, and—

She's got up from the coughing, not wheezing.

All right. Now, have you also wheezed during this period?
What period.
This period of the last week and a half or two weeks.
Yes! Yes!
Do you wheeze during the day?
(Long sigh) I wheeze maybe— I— for a period.
Minutes period, hours period.
A period—I, yeah—I—oh yeah. When I— when I start wheezing, you know, when I— (cough, cough, wheeze) when I have— because then I have a little trouble, you know, with my breathing.
Mm-hm.
It's a good thing I'm not, you know, trying to sing, at the moment.
Mm-hm.
When I do, I take what I have been taking for— to control hayfever, which is Chlor-Tri— ah, no, Pyribenzamine.
And does that help your wheezing?
It see— I think it does, yes.
Right, now when—for how long (I'm now talking days) for how many days, has it been a good thing you're not still singing?
Oh, I would say—a month!
Mm-hm. Now when you were in South America, did you have any wheezing there?
No. No. We were up at twelve thousand feet.

The patient seems to be twisting and turning to avoid being caught by questions that would reveal how restricted his breathing has become. The physician, however, is equally intent on obtaining the information. Sometimes, when I keep at patients in this manner, they will say, "You sound like the district attorney." It may be useful to explain that nobody knows more than they do about their illnesses, and that the questions are merely to bring their knowledge to light.

This next patient was presented in the previous chapter, but without the physician's questions. See how her story acquires diagnostic value when the questions are present.

But before I left, I think the very same— the very next day, still there, I was just sitting at the table and suddenly I got a very bad pain in the back of my uh, left leg. . . . About over here.
Mm-hm.
As, uh, if maybe a muscle were pulled, or something like that—just ached. A Charlie horse. And then that developed so that within two

days I could barely stand on that leg. I didn't know whether— I figured it was a muscle, I thought for a moment maybe it's something vascular—but this is— my other leg is the bad leg, from the vascular point of view. I had thrombophlebitis in it. This, that was my good leg.

Mm-hm.

Anyway, that lasted for about two days and if I stepped on a— a stair going up or down, it was very painful.

Mm-hm.

And then that started to subside. But before that subsided, I started getting a sense of pain, um, in two other places. One was abdominally—not cramps, really, but just a pain. And another was in the middle of the back. I don't remember which one came first. But the whole procedure has been one thing following another.

What was the pain like in the middle of the back?

Um, well it was— it was a— at the beginning these were all like points of pain over a very small area. Very localized.

Yeah, but what KIND of a pain? Was it sharp, dull, achey, burning?

Um, well, I think it was pressure.

Mm-hm.

And it wasn't too terrible except that I was aware of it.

And did you know what made it worse? Did lying down make it worse, did breathing make it worse, did clothing—

Well, at that point it didn't bother me especially.

Mm-hm.

It was just the one spot.

Right.

And it wasn't too terrible.

Mm-hm.

But I was aware that something was happening.

Right.

And then after that, um—

Where are we in time now, would you say?

I'd say these— these things were within two or three days of each other.

Mm-hm. And you said, "a few weeks ago" you were in Rochester—

Oh, I was there November 2nd—

Mm-hm.

—then for several days after

So we're still talking about into the week of November 4th.

Yes. From approximately November 1st through—

Mm-hm.

I returned home on Wednesday of that week. Wednesday or Thursday, I think.

That would be the 6th.

I returned back to New York.

Mm-hm. And you had these symptoms then?

I ha— no. In Rochester I had just the eye and the leg. I came home with the bad leg.

Mm-hm.

I had no trouble, especially in— in driving, or anything like that.

Mm-hm.

And I got the thing in the back and occasionally—but then I got something in the— I'm wearing a rib support. Um, I got an ache over here, also very localized and that was sharp and ah—

Did that hurt you when you took a deep breath?

Well, that's excruciating. I still have that.

When you take a deep breath it's excruciating?

Well, it was— I've been negl— Let me tell you this story. I got that there, and then I got something also in my neck here. After that I got, um, a kind of a dull headache over my eye. Everything on the left side.

Please. Can I go back. The pain that you developed here and just below the ribs, or at the ribs—that was the same character pain that you had back there?

Well, I think at the beginning it was.

But it developed into a more excruciating pain?

It developed into a more excruciating pain.

And what made it excruciating? Its severity or its constancy or its nature?

Well, I think its nature. I'd— what happened— I had at one time a broken rib. I couldn't remember which side it was on, but it appears it was on that side, because I did see—

Did this feel like that?

Well, it felt like that to the extent that if I turned, it hurt a great deal.

The pain that she describes in her left calf does not suggest a "pulled muscle" to me. If the muscle injury occurred during a game, or while running, it might be worse the next day, but it would not continue to worsen. If the injury happened as she said, it should start to feel better the next day. Since most of us have suffered from a pulled muscle, we have experience against which to test the patient's statement. Even in the absence of a complex history such as this, progressive pain following a simple injury should raise other possibilities, such as hematoma or compartment syndromes, which might require immediate attention. The point is not that one can know with certainty what the problem is but that one can get an idea of the degree of *uncertainty* and of how urgent it is that more information be obtained.

The symptom of pain is so important that physicians have developed a large descriptive vocabulary. Laypersons often lack a thesaurus for pain ("It's a pain, how do you describe a pain?"), so the doctor must supply the words with which the patient can compare the sensation. Pain has certain characteristics which often allow its source to be known with great accuracy:

Character. Is it burning, sharp, dull, achey, knifelike, needlelike, throbbing, pressing, squeezing, boring, tooth-achey, gnawing? Given descriptive terms, the patient will usually be able to pick an appropriate adjective. Metaphor is invaluable. "Is the feeling in your feet like walking on sand, or like walking on a very thick carpet?" "Does it feel like a rock in your stomach?" It does not matter whether the metaphor describes something true to life: "Does it feel as if a balloon were being blown up in your finger?"

Constancy. Does it come and go in waves, or intermittently, or does it never let up? This aspect of pain is so important that your questions should continue until you have all the information you need.

Severity. Is it annoying, agonizing, unendurable, discomfiting, awful? As noted previously, the way the patient uses language for other purposes is an aid in evaluating the description of severity. What was the patient *not* able to do because of the pain—work, walk, eat, have sex, sleep, sit still, lie in one position? To what in the patient's experience can this pain be compared?

Duration. Does it last hours, minutes, days, months? Has it been there for years?

Onset. Exactly when did the pain start? Did it begin gradually and build? Did the patient become aware of the pain without being sure when it started? It is useful to place the patient in the precise time, place, and activity where the pain started. Avoid accepting generalities such as "I was home." In what room, doing what, with whom? Before lunch, on first arising, walking to the shower, in the shower? The more accurately the symptom can be placed in the day's events, the more information about the symptom will be forthcoming.

Recurrences. Is this the first episode? "*Never* before, not even ten years ago?" (It is difficult to underestimate the comfort a physician can draw from knowing that a serious sounding symptom happened in the past without serious consequences. To protect oneself from false security, the next question must be "Are you *sure* it was the same?") What were the circumstances of the recurrences? If there have been many episodes, one wants a description of the most severe and then the least severe, and finally, some sense of the distribution of attacks between those extremes.

Associated symptoms. Was there nausea, vomiting, anorexia, diarrhea constipation, fever, weakness, sweatiness, fear, hunger, happiness, sexual excitement, anger, or even other pains? Symptoms *not* associated are also important. For example, abdominal pain with *no* associated abdominal symptoms, such as a change in bowel habits, loss of appetite, nausea, fullness, or relief that comes from belching or passing gas, makes one wonder whether the pain actually arises in the abdomen. Perhaps it is referred pain, in which case other associated symptoms, such as back pain, should be sought.

What makes it worse, what makes it better? Is it aggravated by food, position, activity, emotion, work, play, and is it relieved by any of those? What does the patient do to obtain comfort? Many of these questions can be asked in such a way that the patient conducts part of the physical examination. "Sit up straight, and put your left arm straight out in front of you; push out as hard as you can. Does that bring on the pain?" "Take a deep breath. Did that make the pain worse? Better?" "Raise your arm straight over your head. Can you do that?" "Touch your chin to your chest. Does that hurt?" "Press on your chest where it hurts you. Does that cause you pain?" As you can see, these questions can be asked by telephone.

What did you do for it? Was the pain relieved by aspirin, a hot shower, antacids, Alka Seltzer, a heating pad, codeine, or other treatments or medications?

What have other doctors said? If the pain has been present for a prolonged period, or has recurred frequently, other physicians have probably been consulted. What did they say, what tests did they perform, what treatments were prescribed? If you are lucky, somebody else has done all the work.

Rarely does one have to proceed through the entire litany before recognizing the pain. Occasionally, however, one has to pursue each of these points like a tiger after a gazelle; as with the tiger, the effort will be rewarded. It is probably safe to say that more diagnoses are missed because of failure to delineate completely a symptom than because of any other omission. As the years of practice go on, most members of the family of pains become easily recognizable old friends. From time to time, however, a challenging, perhaps life threatening, new pain is described. When this happens, the full range of questions will again be necessary. What has been said of pain is true of most other symptoms. I believe a beginner should memorize the pain questionnaire and repeat it in front of the mirror until it is forever locked away. Learning these questions is worth at least as much time as studying the origins and insertions of the muscles. Indeed, the two kinds of knowledge working together makes for good clinicians.

Note how much patience and persistence is necessary in pursuing the symptoms of this next patient. The patient has the data, and the doctor means to find out no matter how the patient twists and turns to avoid answering.

And I got the thing in the back and occasionally—but then I got something in the— I'm wearing a rib support. Um, I got an ache over here, also very localized and that was sharp and ah—

Did that hurt you when you took a deep breath?

Well, that's excruciating. I still have that.

When you take a deep breath it's excruciating?

Well, it was— I've been negl— Let me tell you this story. I got that there, and then I got something also in my neck here. After that I got, um, a kind of a dull headache over my eye. Everything on the left side.

Please. Can I go back. The pain that you developed here and just below the ribs, or at the ribs—that was the same character pain that you had back there?

Well, I think at the beginning it was.

But it developed into a more excruciating pain?

It developed into a more excruciating pain.

And what made it excruciating? Its severity, or its constancy, or its nature?

Well, I think its nature. I'd— what happened— I had at one time a broken rib. I couldn't remember which side it was on, but it appears it was on that side, because I did see—

Did this feel like that?

Well, it felt like that to the extent that if I turned, it hurt a great deal.

Did it hurt— How about when you breathed?

Well, I can't—yes.

Then. Then.

Well—

Did it hurt when you breathed then?

Well, it took—at the beginning it wasn't that bad, no.

After a day?

Yes, it began to develop.

Okay.

I'd say that all these—from the moment I was first aware of the pain, 'til—it took about two days 'til it developed, and sometimes I could tell where I was going to get it.

Mm-hm. How could you tell?

By just having a slight sensation.

I see. We are now in what week? You came back on the 6th, that Wednesday. Had you had this thing, this thing now, by the weekend or the following week?

Do you have a calendar?

I sure do.

We're up at— Today we're at the 22nd, aren't we?

Yep.

Well, I would say we're at the weekend of about the 9th, where I was feeling quite—

Right.

—miserable.

Next question. Had you seen a doctor yet with all these pains?

No, I kept putting it off. I'll tell you why.

OK. Did you have fever? I don't care, I don't care why, to tell you the truth. Did you have fever? I do, but I don't at the moment. Did you have fever?

I don't think so.

Did you feel otherwise sick.

I felt hot and— yes, I felt— no, no, just not right. Maybe hot and cold.

Did you feel sick?

Maybe feverish, but I didn't— I don't think I had a fever.

Did you feel sick.

A little bit, yes.

There we have the classic story of thrombophlebitis followed by pulmonary embolism. The clarity of the correlation between whole human pathophysiology and symptoms can be gratifying. The belief that the patient has a pulmonary embolus comes from the history of pain in the leg followed by pleuritic chest pain. The chest pain alone would alert us to the possibility of embolism. Pleuritic pain is a special variety of pain. As the roughened pleural surfaces rub against each other with each breath, a thin stabbing knife of pain causes the affected chest to "grab" and splint, stopping the breath. It may make the patient feel short of breath because of the inability to breath deeply; it is extremely fatiguing because the patient cannot escape from the pain. Blocking the appropriate intercostal nerve(s) is a simple method for effectively relieving the distress. It is amazing how five or ten milliliters of lidocaine can transform a patient in agony into someone who is grateful and very cooperative about giving a history, even in the middle of the night. Interestingly, even when the lidocaine wears off, the pain usually does not return as severely. I suspect this occurs because the intercostal muscle spasm, which becomes a part of the picture of pleuritic pain, usually does not recur. Since muscle spasm is a part of the story, it is possible to mistake the pain arising from spasm in other muscles of the chest for pleuritic pain. Spasm in the rhomboids (or the trapezius) will produce lancinating pain in the posterior chest, which occasionally radiates to the ipsilateral anterolateral or anterior chest wall and is aggravated by breathing. Motion of the scapula produced by hunching the back or thrusting the ipsilateral arm forward intensifies the pain of the muscle spasm, but not the pleuritic pain. In addition the muscle itself is tender where the pain arises. Appropriate questioning can often, but not always, distinguish the two. This is important, because the patient's dire interpretation of the pain of muscle spasm can influence the report and thus can be misleading.

Why it took this patient so long to do something about her pain is another part of her illness, to which we will return later.

Establishing the Time Line

In the previous history, as in most, establishing the temporal sequence is difficult but vital. The trouble stems from the fact that, in obtaining a history, physicians are interested in real time—the clock and calender variety—whereas patients internalize a subjectively skewed version. The place where events transpired is usually not as crucial in stories of illness, but since place and time are frequently related in the memory, it is useful to work at the two together. Keep in mind that almost no matter what a trial the process of discovery may be, somewhere within the patient is stored the information you are after. That inner storehouse of information (which I fancifully termed the *experiencer*) requires some assistance, however. The patient is an unwitting victim of the tricks of memory and the distortions of past experience, which obscure the actual train of events. What eclipses the objective chain of events are the meanings the patient has assigned to these events and the causal nexus into which they have been placed. To help the doctor's friend, the experiencer, one must provide an external framework of time and place and separate the events from the patient's beliefs about their significance and importance.

In seeking the time of onset of long-term illness, it is useful to provide specific temporal landmarks. "Were you coughing during the Christmas holidays last year?" "No, I'm sure I wasn't." "Were you coughing during Easter recess?" "I think I was, but I'm not sure." "Where did you spend your Easter holiday?" "With my friend, Sally, at her home in Cleveland." "What was the weather like?" "Now I remember: the weather was cold and terrible, and because I was coughing so much, we practically never went outdoors." "So the cough started sometime between Christmas and Easter. What courses did you take last year?" And so on. The thrust of the questions puts the patient in a *specific place*, doing *specific things*, at a *specific time*. One is merely taking advantage of the associational character of memory. Vacations and holidays are useful because they stand out from the day-to-day past. In short-term illness the time of day, place of work, mode of transportation, weather, and many other things provide the associations with which to bring out the facts.

Generally, people underestimate the length of an illness. The recurrent asthma, which was originally believed to have been of two years duration, turns out to be four years old! Therefore one should not accept an onset without going on to establish when the patient was previously entirely well. Just as some illnesses are underestimated, others are overestimated, particularly if the person is using the illness as evidence for some belief: "I've had one cold after another for months; I'm sure I've got a basic infection in my system because I'm so run down." With persistent questioning, the number of "colds" turns out to be two, and the "months" turn out to be weeks.

Contrast this next history with what has been presented thus far. The rationale for the amount of detail about history-taking methods contained in this chapter and the next is to avoid the errors committed in the next example. The next patient is talking with an experienced physician who is taking a poor history. An inadequate history, though it may take less time, is not an economical use of time; it is ultimately wasteful because it leads to wasted effort. This doctor is not trying to find out what happened; he is out after a diagnosis.

So how are you feeling, Miss——?
Well—

Feeling good for the last few months?
Well, actually a couple of weeks ago—

Yeah? What happened?
Well, I ran into a bad case of diverticulitis. I know I've had it.
Mm-hm.
But when I was here for the different tests a year ago—
Mm-hm.
I had been well and without it for so long that it slipped my mind—
Mm-hm.
—completely.
That's the way it stays, dear. It comes and goes.
Ah, yes. Now. I really didn't call up for that. I called up for some other—
Go ahead.
Anyway, I was forced to go to a doctor across the street,
Mm-hm.
because my temperature was 102. And I had—
You had trouble with your diverticulitis?
Oh yes! And how! So I was in bed most of the time—
Did you have diarrhea or anything at the same time?
Did I—? Excuse me?
Did you have diarrhea? Loose stools?
To an extent, yes, but— to an extent.
And you had pain?
Pain on this side, yes.
Sounds like diverticulitis again.

The question, "Did you have diarrhea or anything at the same time?" is what is often referred to when leading questions are discussed. It is not the question that is poor but rather the follow-up. The patient answers "To an extent, yes." What does that mean? Yet the patient's affirmation seems important to the doctor's conclusion.

It would be wrong to give the impression that the true chain of bodily events is always like a vein of gold buried deep in the mountain. Many, perhaps most, histories are straightforward.

Here is a straightforward story.

Yes Ma'am. How can I help you?
I have a pain in my stomach. And, um, I wake up during the night, and it's quite a severe pain.
Wakes you up at night?
Yes it does.

How long has that been?

Well—

How long have you had the pain altogether?

Altogether? I would say, almost a year now. Well, it was not THAT bad, you know—

Mm-hm.

I mean, it was—

What kind of a pain is it?

Well, at the beginning it was just like a burning pain. It was, you know. It would start around in the afternoon. I would notice it more, around three, four o'clock. It would start burning a little bit, and then I would start burping. And ah, it would go on for two or three days, then go away, then come back again.

When you ate would it make it better?

Well it was not—when I ate, yes, it would— it would disappear for a while, yes, it was better if I ate.

And how long did it stay like that? That is, there every couple of days?

Oh, it stayed for— for two or three months, maybe more than that—

Mm-hm.

—you know. And then, I don't really— I cannot really say I really noticed how long, you know. But in the past— since Christmas time or so, ah, I have this more severe pain, and it comes in the morning. It comes at different times. And if I even, you know, sometimes I—— I'll eat a piece of candy or a—

Mm-hm.

—cracker or something, you know. It eases it for a little bit but not really too much. But it's not every day, you know.

How often does it wake you up at night?

Well, the waking up at night is quite more ofen.

Mm-hm. How long has that been?

Oh, for about three months.

What would you do for it when you get up in the middle of the night?

Well, I got some of those antiacid pills.

Do they help it?

Yes, they help it. I suppose I fall asleep again after, so I suppose it helped.

Do you take those during the day also?

Occasionally I do.

Do they help then, too?

Yes! I would say they did. But, you know, this— it's still uncomfortable, and I feel something has to be done with it. It's not—

Does it stay in the one place or does not go—

No, it's— it's in one place.

Is it always the same KIND of pain? Although maybe different in severity?

Sometimes it's— you know, sometimes it's just like a burning—

Mm-hm.

—uh, sensation. But then other times it's a real pain.

Mm-hm.

And—

Does it ever make you nauseated?

Yes.

And how long has that been that it gets you nauseated?

Well, I don't really know HOW long, but I've been getting nauseous more often.

Do you ever get nauseated when you don't have the pain?

Yes, I do.

In the beginning did you get nauseated even without the pain or in the beginning was it with the pain? In other words, did they come together, or were they separate symptoms, do you think?

I don't know.

Have you vomited with this pain at all?

No.

Or vomited during this period of time?

No. I only did once, but I feel that I— I must have eaten something. But otherwise I didn't—I don't, ah—

What seems to make your pain worse?

I don't know!

Alcohol?

No.

If you take a drink, does that make the pain worse?

No. No.

Do you smoke cigarettes?

No. I don't smoke. I found— Well, I tell you I thought. I thought that if I would, if I would eat something bland, you know,

Mm-hm.

It would ease the pain.

And does it?

Say, if I would have it during the afternoon, at night when I go home, I thought, if I would have some uh, maybe noodles with cheese or something, no meat, no sauce or anything, but that doesn't help. That makes it worse. If it—

How long after you eat does it get worse?
Oh, it— quite shortly. I would say maybe, within an hour—
Mm-hm.
—it starts burning, and—
Now, wait. Does it go away and come back?
Yes.

The patient did not have an X-ray proven duodenal ulcer! But the patient did not describe a "duodenal ulcer"; that is a doctor's word. She merely provided a verbal representation of excess gastric acid secretion. In some with this history a duodenal ulcer can be demonstrated, but not in others. Medical statistics are usually provided in the form, "X percent of patients with demonstrable duodenal ulcer have night pain." While that may be important, it would be more useful for clinicians to know what proportion of patients with night pain (or the other features of her history) have demonstrable duodenal ulcers. Negative X rays, or even endoscopy, would not contradict her history, nor change her immediate treatment. No one doubts the comfort provided by the objectivity of an X ray, but the patient's history is also part of the doctor's objective knowledge, in the same manner as a heart murmur is objective. The doctor hears both, and both stand in some probabilistic relationship to a pathophysiological process. Perhaps as training in history taking improves, physicians will come to trust what the patient says as much as what the X ray says, thus become aware of the fallibilities of both sources of information.

The next patient presents one of the most frequent symptoms women report to physicians—fatigue. This symptom is often disregarded by physicians because it is too nonspecific to be a reliable indicator of disease. One can understand medical disinterest, because fatigue can result from working hard, boredom, depression, inadequate sleep, inadequate food, as well as from hypothyroidism, Addison's disease, and chronic illnesses of many varieties. The perception of fatigue, however, arises from a source of information that is generally extremely reliable—the person's general state of being in relation to his or her world. One knows whether or not one feels tired. Is there a contradiction, then, between the reliability of the report of fatigue and physicians' perception of it as an unreliable indicator of disease? There is no contradiction. Patients go to doctors because they think something is wrong, and doctors are primarily interested in troubles that fall within the classification of disease. This is unfortunate on two counts. Doctors who brush aside fatigue, often by saying "It's nerves" (whatever that means), miss the chance to make patients feel better, since we doctors often know how to make fatigue go away. We may also miss the chance to make early diagnoses, because fatigue is commonly an initial symptom of disease.

Generally, "normal" fatigue and most disease-related fatigue is directly related to energy expenditure, whereas that related to emotional causes is not tied solely to energy output; for example, the fatigue of emotional origin might occur with office work but not a tennis match. Therefore the primary task in delineating the symptom of fatigue is finding its relationship to energy expenditure. In this next example, this process can be seen.

Well, what can I do for you?
Um, I need a checkup, but more for— I'm just not feeling well. I haven't been feeling well for months.
What kind of not feeling well?
Tired, nauseous—
And when do you get tired and nauseous?
Right now. (Laugh)
And when will it come on?
Ah, usually d-during work.
You wake up feeling all right?
Yeah. Usually.
And then when— when do you start getting very tired?
Sometimes as early as 11:00.
Mm-hm. What kind of work do you do?
Teacher.
When do you start getting nauseated?
Um, seems to be after I'm ti— feeling very tired.
For how long. In other words, does the nausea and the fatigue start at the same time?
No. I'm usually tired, and I get nauseous.
Before lunch or after lunch?
Usually I don't have lunch.
When— You don't have lunch.
No.
Do you have breakfast?
Yeah.
What do you have for breakfast?

Cereal.

And?

That's all. Coffee, juice.

Well, do you have coffee and juice, or are you adding that for my benefit?

No. I have coffee and juice. Sometimes I have cereal, sometimes I don't.

The cereal was for my benefit.

(Laughs) Right. Um, yeah. Depending. Like sometimes it's— for the past month I've been taking— I can't get through the day without taking like a three- or four-hour nap.

When?

By the time I get home around three-thirty or four o'clock.

And then when you awaken from that are you well again?

Yeah. I feel pretty good. And then I get tired again by around eleven.

Mm-hm. And how long's this been going on?

About a month—the sleeping and taking naps.

And how long's the whole fatigue?

Ah, since October.

Since October. Can you date it exactly?

(Sigh)

Do you go out at night?

No. (Laugh)

When did you stop? Were you dating before this all started?

Um, I guess it really started while I was working; about a month later.

It started when you started working, you mean?

Yeah, I work pretty hard but—I've done teaching—

Is this the first year you've ever taught?

No, I've taught before. Last year I was a full-time graduate student, and I didn't— I started teaching in May. I got a job with——

Were you tired like this in May?

No. I had twice as much work because I was still in school—on a full-time program.

Mm-hm. What did you do during the summer?

I worked most of the summer as a waitress.

Mm-hm.

It was exhausting, too!

Mm-hm. Well, were you tired when you worked as a waitress? . . . You weren't tired like this then.

Mm.

You started school, when? What month?

Ah, September.
Do you have a regular boyfriend?
Yes.
How long have you had him? Since you started teaching?
Since June.
Since June.
Since he's been living with me.
He lives with you. All right. Now, what time do you go to bed nowadays?
Eleven, eleven-thirty.
And when you've wakened after your nap and then you're all right until about eleven/eleven-thirty.
Mm-hm.
Did you used to go out in the evening?
Before I started working?
Mm-hm.
Um, occasionally. I really haven't had the money to go out. (Laugh) Probably more often, but like, for the month of August I wasn't working,
Mm-hm.
So I probably went to more movies then.
Mm-hm.
You know, I'm not a real swinger, in that sense.
Do you get tired like this on weekends?
Ah, Friday evenings I'm usually conked out. Saturday—
How about Saturday?
Saturday I could still go to sleep by twelve.
By twelve what? Noon?
No. By twelve o'clock at night, which is—
Have you taken a nap on Saturday?
No. No. But then I've slept until about eleven then.
Mm-hm. And on Sunday?
Sunday? Sunday's kind of— a quiet day.
What do you mean?
You know, *New York Times*, Lower East Side, come back, you know, watch TV—
Are you tired like this on Sunday?
No. No. No, I feel pretty well-rested. But I haven't exerted myself that much either.
Mm-hm.
A few times— a couple weekends I remember dragging. The past

couple months I haven't. But then again I was, like, sick a lot during the week.

And it's worse in the last month.

So before the weekend I had, like, a good two-day rest, so I really couldn't have been tired.

This is the story of organic fatigue in a patient with previously undiagnosed hypothyroidism. My initial question is usually, "Are you tired when you wake up in the morning?" If the patient awakens fatigued, there is a good chance that nonorganic factors are at play. Patients such as this one are very grateful when they discover that there exists a physical cause, and even more so when the solution to the problem is simple. Reports of fatigue, like those of other symptoms of the general state of being, must always be believed until proved false. To repeat what I have said before, when a patient says that something is wrong, then something must be wrong. If your arm does not work in the familiar manner, you know that, and so it is with every part of you. Because patients are frequently incorrect in their attribution of cause, their doctors may dismiss their entire history. Theirs is not to find the cause; that is your job. It is sad when the patient's symptoms are discarded for the foolish reason that, on first listening, the doctor disbelieves them, or for the callow reason that he or she has never heard anything like that before. There is a concreteness to organic symptoms, however, that gives one confidence in their organic basis. As odd as they may, on occasion, sound, they usually have internal consistency, although this may require searching out. For example, the dysfunction occurs after similar kinds of provocation, or in similar situations, even though a "cause" is not evident. Symptom reports are the same from one visit to the next. If the description of the symptom is written down, four months later the patient's description will be exactly the same. When this occurs, chances are that something is wrong with that person's body, and it behooves the doctor to find out what. If, unable to make a diagnosis, you pay sufficient attention to understanding how the symptom disables, you may be able to make the patient better. Such patients have usually paid many futile visits to doctors before you; therefore after their initial distrust of physicians is ovecome, you will have very cooperative allies. Their distrust has been earned. It will melt away in front of your concern, however, which is demonstrated by your continued attempt to understand their symptoms rather than merely to force the symptoms into some diagnostic box.

But suppose that, after careful listening, you believe the fatigue, or any other symptom, is psychogenic. In this case you are no less required to support the conclusion with sound reasons. *"Psychogenic" should never be a diagnosis by exclusion.* Either find out what emotional event(s) brought on the tiredness or find other specific evidence to support your impression of an emotional problem that is *temporally related* to the onset of the fatigue. One should not accept statements such as "I've been under a lot of stress lately." For one thing, so has everyone. If the stress is sufficient to change the person's self-perception, however, a concrete description should be possible. It is probably fair to say that when a symptom description resembles cotton candy—looking good but melting into nothing—then the pertinent facts are not yet out. If one believes that depression underlies the fatigue, then other evidence of depression should be sought. Without supporting evidence it is, at the very least, unfair to suggest that emotional difficulties are present. Moreover one should no more make an unsupported diagnosis of psychological problems than of, say, pheochromocytoma. Some action follows virtually every diagnosis, even if it is only self-doubt on the part of the patient. To the degree possible, every action of the physician, and every action which the physician promotes in the patient, should be based on supportable evidence. Just as one carefully builds the case in support of the hypothesis that organic disease is present, so should one proceed with nonorganic illness.

If I suspect that fatigue, or any other symptom, has a psychic origin, I am initially circumspect in my search for the personal or emotional elements to which it is related—more cautious than in the search for the symptoms of organic disease. This is because of the following factors. First, patients may feel abashed, or even ashamed, to discover that their abdominal pain is related to conflicts on the job. Therefore they may defensively deny something which is actually the case. (Remember, one never wants to push someone into telling an untruth; recovery may be difficult.) Second, when the reason for the symptom is completely apparent to patients, they are quick to tell, as in "Oh, that's just my nervous stomach. I always get diarrhea when I'm working on a lecture; I was hoping you'd give me some Librium." Allowing for the satisfaction of my diagnostic scepticism, Librium it is. But when the patient is not aware of the personal or emotional source of the symptoms, the necessary questions will seem like personal intrusions, or just plain nosiness (a behavior to which no one, not even a doctor, is entitled). Third, sometimes the patient suspects that the illness has a psychological basis but is hoping that the doctor will find something that a pill will cure. In these instances patients may be less than candid about their personal lives. When I begin to hear evasiveness, or when a frontal assault on the facts seems unwise, I insert a small question here and, another a few minutes later, another while I walk with the patient to the laboratory, several more as part of small talk while the patient gets adjusted on the examining table, another between examining the fundi and looking into the throat, and so on. True, I am being sneaky. These questions are part of my train of thought, but I am attempting to hide my thinking from the patient by asking questions that seem innocuous because I am obscuring their interrelations. Indeed, when it is well done, it merely sounds as if I am personally interested in the patient; since that is also true, this manner of questioning serves two functions. When the patient and I sit down together after everything else is finished, and I am ready to share my hypothesis about the nature of the patient's problem (and, ultimately, what I believe should be done about it), then the evidence that I lay out will include the information obtained through the oblique questioning. If more is needed at this time, I can be more forthright in asking—*particularly since no matter how the information was obtained, the patient is not being accused, either of the facts, or of withholding the facts.* I cannot stress this too forcefully. Remember, no matter how negatively a patient behaves, part of that patient must have wished you to discover the facts in order to help, or the patient would not have come to see you! (I am excluding from this discussion patients who have been forced to come to a doctor, or where a doctor's skills are being used with persons outside the context of a normal doctor-patient relationship—to determine disability, and so on. I believe that physicians in these situations are bound by ethical constraints *not* to use their special skills *against* any person.)

The foregoing case exemplifies how much can be learned by asking questions about daily living. In fact, obtaining a description of someone's day in exhaustive detail will tell a remarkable amount about the person. It may take between forty-five and ninety minutes, and I suggest it to you, even if only as an exercise with a willing subject. On the other hand, it takes only a few minutes to discover that the patient's fatigue is a result of the side effects of medication, insufficient sleep, an exhausting job, not enough food or exercise, being overwhelmed by too much responsibility, and many other factors in daily life that are well known to doctors and barely part of many persons' knowledge. Doctors know those things because we have learned (if we listened) about how people live, and patients are usually grateful for the advice about how to feel well again.

The next patient was presented in the previous chapter. Again, we will see how the history-taking interaction changed the focus of the physician's attention.

Now, in August you were throwing up and in September, too?
Yeah. I mean, it stopped.
Mm-hm. Well, you're saying it stopped for a while. When did it stop?
Yeah. Ah, I had it in late August, I just had it for, maybe, two days, and then it stopped.
Mm-hm. And then when did it come back?
Um, Let me think—think it um, came back in October. I think I went through September without having it.
Mm-hm.
It came back in October.
I see. And when it came back in October, was it frequent then like it is now? Did you have a week or so?
No.
No. Mm-hm. But this week it's been frequent.
Mm-hm.
Now. So, August, then it wasn't there in September, then October it was there but it wasn't bad like it is this week?
Right. I wasn't— I didn't bloat up. That's what really bothers me.
Now, I'm on vomiting now. I'm on vomiting. You mean if you didn't bloat up now you wouldn't mind vomiting?
Not that much. I would—yeah, actually I would.
Mm-hm. I mean, but if it was just vomiting, would you be here? . . . . .
Probably not.
I see. Okay. Now. Do you vomit on weekends?
Uh-huh.
And you vomit on weekdays?
Mm-hm.
And you usually vomit at lunchtime? When you vomit?
Yeah.
And do you get rid of all the food you've eaten?
No.
No.
I— I'll throw up, you know, once. And then it'll come on me again and I'll throw up again. But usually I don't get everything out of me.
Mm-hm. Okay. And then you started to get this bloatey feeling. When was that?
A week ago.

And has it left you since then?

Hm-mm.

What does it feel like?

It feels like, um, my stomach's just blown up with something. Something's in there, you know. It kind of hurts when I poke it around.

Mm-hm. And is it swollen, too? Are your clothing tight on it? Are your pants tight?

No, not really because I—

Does it look bigger to you?

Yeah, feels bigger.

And do you belch a lot?

Uh-huh.

And is that something new for you?

Mm-hm.

And when did you start?

Um, a couple of days ago.

Mm-hm. And do you pass gas?

Mm-hm.

Do you pass a lot of gas?

Well, I guess a lot compared to what I did before.

That's what I mean.

Yeah.

Mm-hm. And when you pass gas or you belch, does your stomach feel better?

Mm-hm. Yes. A lot better.

And then how long before it comes back?

Well, it— it never feels altogether better. But it'll come back in two hours maybe?

Mm-hm.

And I wake up with it. I wake up with my stomach more blown up.

Blown up again. Mm-hm. And you've been constipated this week, too?

Yeah.

Is that something new for you also?

Yeah. I don't usually get constipated.

And when you mean constipated, what do you mean by that.

That I can't go to the bathroom.

And when did you last go to the bathroom?

On my own?

Yes.

Um, four days ago, maybe?

Mm-hm. And when you say, "on your own", what do you mean by that?

Without enemas or anything like that.

Do you generally use enemas to go to the bathroom?

Well, yes, since—

Since when?

Since um, about—about a week.

And is that the first time you've ever used an enema?

No.

When did you start using enemas?

When I had this in August , I was constipated but I wasn't blown up, and I used enemas then and it just went away.

Mm-hm.

I don't know why, really.

Mm-hm. And how many enemas would you say you've used since August? A hundred?

Guess—yeah. A LOT. A lot.

And what kind of enemas do you use?

Fleet enemas.

Well—if you've used a hundred enemas, you must have been constipated for more than just a week now and then a little time in August, right?

I— well, it seems like a lot because I use a lot of them. I mean, in August I was, and I wasn't for September, and then—see, I don't just use one enema.

Mm-hm.

I use one, and then I use another one, and sometimes—

Why is that?

Because there's still something—I feel like there's still something in there, and it's up higher, you know?

Mm-hm.

And I try to get it.

I see. So it isn't just that you use the enema but that you try to get something out that you don't think is getting out all the time?

Yeah. Yeah.

And has that also happened in October, too? Even when you aren't bloated?

Yes. It did.

Mm-hm. And who told you about Fleet's enemas?

Who t—? My mother.

Mm-hm. Right. So you go out, you get yourself some Fleet's enemas and then— how many enemas will it take before your, ah, your bowel's empty? Finally. And you get it out.

Well, I never really get it all out. That's how I feel.

Mm-hm.

I never get it all out. But usually, I'll take, mm, two, and that's about—it'll all get out. All that's going to come out, and then I keep trying because I think maybe some more—

And how many is the most enemas you can remember using over a day?

Over a day—six, maybe?

Mm-hm.

What a different case it becomes. It is somewhat disturbing to hear this story from a fifteen-year-old. We will hear more from this patient in chapter 4. Again exemplified is the pursuit of the details, until the doctor understands "what brought her to the doctor." Also clear is the need to listen to every word. "When you mean constipated, what do you mean by that?" "That I can't go to the bathroom." "And when did you last go the bathroom?" "On my own?" Following up on that utterance brings the physician to the problem.

"And how many enemas would you say you've used since August, a hundred?" This form of question does two things, it gives patients "permission" to have been behaving in an extreme way, at the same time suggesting that their behavior could be "worse" than it actually is. Thus the truth becomes a defense against the unspoken criticism. "Well, do you drink a quart, or two, of whiskey a day?" "Oh, never. Maybe a fifth or a little more."

The next example comes from an admission history taken by a third-year medical student. The student had been taught to let the patient tell the story without interruptions for five minutes or more. Five minutes is a very long time—much longer than is usually necessary to present the basic problem, and also longer than is usual in spontaneous conversation.

Now, tell me what is it that's brought you to the hospital? What is—

Well, it goes so far back it seems like a couple of years. It started eight months ago.

Mm-hm.

With indigestion. Cramps and indigestion. I get it after lunch hour a lot. And, uh, after two weeks of persistency I checked it out with my HIP plan, you know. I happen to belong to that. I'm a city employee, you know. And I was with them for two and a half months, and they were puttin' me through a GI series—barium enemas and checking me out that way, and everything was coming out fine. So they couldn't get at the problem. Now, what really was bad was at night when I'd get into the bed, I'd start getting a push at me, which—which made me get out of my bed. I couldn't stay in bed, and I was up all night. I'd have to sit on a sofa, I'd have to stay up all night—I couldn't stay in bed. I'd have— in some ways it would be gas rippin' through me, in some ways it would be cramps—it would be all just enough to keep me awake through the night, and I couldn't stay in the bed because of this push that was coming against me. And I explained that to these doctors, and they, as I say, they gave me the GI and the barium enemas, and everything was showing up fine. So nobody was getting at the problem. I stayed with them for two and a half months, and I wasn't getting anywhere. One morning my wife saw the condition I was in, when I— she woke up—her doctor is a doctor here.

Mm-hm.

And she called him up and explained the condition I was in,

What condition was that, that you were in at that time?

Well, I couldn't sleep. This was going on for over— for up to four months time! It was going on for four months! I couldn't stay in the bed! At night—all through the night I was awake. I couldn't sleep. There was all kinds of sickness going on in the stomach, and I had to stay— I couldn't stay in the bed! There was something pressing into me. I ha— I have to get out of the bed around midnight and from there on it's all through the—up until morning sometimes before I could get back in, you know, for a few hours. As I say, she saw that I was getting nowhere, and she decided to call Dr.——.

Even after the two minutes and forty-four seconds that have been included here, we know little of the patient's symptoms. He was found to have carcinoma of the pancreas that, as you saw, went undiagnosed for a prolonged period, because negative X-ray studies were taken as definitely excluding disease. A normal X ray is not a normal patient, it is only a negative X-ray study. While pathology can be represented on X rays, it can only be *present* in patients or their parts.

Too often histories are taken via a third person. I find that so unsatisfactory as to be useless. Yet telephone histories are commonly taken that way. The third person says, "Oh, he's too sick to come to the phone." Not coming to the telephone can be the cause of many an unnecessary trip to the emergency room. No one else can know what you feel, they can only know what you have said. The basic facts become increasingly obscured when the patient has added some meaning to the symptoms, and then the parent, spouse, or friend has added some further meanings (especially if it is felt that the doctor must be made to understand how serious the illness is by exaggerating symptom severity). Here lies the pediatrician's special problem. But, as those who care for adults become expert at reading the patient, pediatricians develop the skills necessary to find out about the parents *and* thus the child through the parents.

Let me end this chapter with an example that once again emphasizes the basic message of this chapter. Through the effective use of questions, the doctor is able to bring out those events in the patient's story that tell what is happening to the body and to unfold the pathophysiology of the illness, sometimes in exquisite detail.

Night on the hill.
Ah, the night on the hill. Well, there's a very, very steep hill—
Mm-hm.
—and as I was going up the hill, I started to feel pain in the center of my—
What did the pain feel like?
It felt like I was pushing the hill with my chest!
Mm-hm.
And I started to get very, very out of breath, and when I got to the top of the hill, I was panting. I couldn't catch my breath. And as soon as I got to the top, the pain stopped, that pushing stopped.
Mm-hm.
And then I got to the elevator wing, and I rested against the elevator and waited until I got upstairs. And finally when I was standing outside the door, my breath started coming back.

(A thirty-eight-year-old woman, otherwise healthy, is walking up the hill, and she gets chest pain and shortness of breath. The pain feels as though she is pushing the hill with her chest; put another way, the pain is "pressing." Which disappeared first, the pain or dyspnea? The pain. The symptoms are new to her. It sounds like more than a cold. A fair hypothesis is that something is wrong with her heart.)

And you felt fine.
No, I didn't feel fine. I was scared. I felt—
Were you scared while you were going up the hill?
Ah, not until I started panting.
Mm-hm. Mm-hm. I see. Have you ever had any discomfort in your chest like that before?
No, never.
Have you ever been short of breath like that before?
I've been short of breath climbing hills before, but nothing like that.
And since that happened, that was when?

That was last Wednesday night.
Since that happened, one week ago, have you had any repetition of that?
Of that incident?
Of that discomfort in your chest or shortness of breath.
No.
You've had no chest pain since then?
Well, I told you on the phone I did, but it was different.
Tell me about that.
I had it in the left breast on Saturday for about seven or eight hours.
Mm-hm.
Every time I tried to take a breath I had sticking pains. This was completely different.
Right. And— and has that gone away?
That went away, yes.

(In a patient in whom, because of both dyspnea and pressing chest discomfort, the possibility of heart disease is raised, a new chest pain occurs. This discomfort sounds like it arises from the pleura, or the lungs via the pleura. A more refined hypothesis includes both heart and pleura with an acute, recent onset—pericarditis. Can evidence that bears on that be found?)

And that discomfort that you had, was that better lying down, leaning forward, standing up—?
That sticking—um,
Mm-hm.
Well, I think I started to tell you on the phone that I found that, um, since I couldn't get deep breaths, I found that moving forward enabled me to catch my breath.
Mm-hm. Were you surprised by that?
Yes. Very. And it— it seems— and I've been doing it quite a bit. When I feel I can't take a really deep breath, and I hav— as soon as I lean forward, I get my breath. I don't know why.
Right. And has that— that feeling that— that leaning forward would make you feel better, has that been true in the last couple of days?
Yes. Yes. It's a feeling of relief. Right.
Is that true today?
Is it true today? I would have to do it to see.
Do! Now, the burning that you had in your chest prior to one week ago. Is that burning relieved by any position? Like leaning forward?
Sometimes when I would lean forward, yes. (Cough)

(Why should leaning forward make her more comfortable? Sometimes people with abdominal pain lean forward because it takes the abdomen out of stretched position. But she does not have abdominal pain. Neither does she squat like children with congenital heart disease. Leaning forward gives the heart more room in the chest, allowing it to pull a bit away from the pericardium. This maneuver frequently produces relief in pericarditis, especially before much fluid accumulates. More information is needed bearing on the hypothesis.)

Have you had any pain in your shoulders or neck? Or in your back since this all started?

Shoulders, neck, back. No.

Mm-hm. And has that funny cough that you get—has that changed at all?

Same thing for three months.

Mm-hm.

Usually when I'm lying down, that's when I feel the irregular beat's stronger. And when it's very irregular, it makes me jump, and that's when I start to cough.

Could you have that cough without irregularity?

Could I?

Mm-hm.

I don't know—I never had it before the— before the irregular heart beat.

You said on the phone that you've been very tired lately.

Extremely.

Is that something new for you? This quality of fatigue?

Very new, yeah.

I mean, you've had times when you were tired—

Yes, yes.

Is this different than that?

Yes. This in not like when I was tired and recovering from the flu in March.

How is it—

And, um, because then my arms and legs were tired—everything was tired.

What's tired when you've tired now?

Um (sigh), well, I get—let me give you an example. This morning I was stripping the beds.

Mm-hm.

And after I stripped three beds I felt exhausted.

Mm-hm.
I felt like taking a deep breath and sitting down and resting.
And did you?
Yes I did! And that's not normal!
And how long did it take— how long did it take before you started to feel better?
Oh, I only sat for two minutes and then I got up and did the—
And then did you feel better?
Yeah.
In other words, this fatigue goes away in a few minutes.
Yeah, if I sit down and rest.
Mm-hm. But a fatigue like with the flu, would that go away in a few minutes?
No. No. No.
Do your legs get tired when you get this fatigue?
No, aching in my legs, no. No. (Cough)
Mm-hm. No fever?
No.

And so it went through the history. The patient had both pericardial and pleural friction rubs and ST changes in the chest leads of her electrocardiogram. Two days later the auscultatory findings were gone, and not long after that her electrocardiogram returned to normal. Other studies were not diagnostic. (Echocardiography was not yet easily available.) Her history, however, never disappeared. Thus the only place from which a physician in the future might be able find out what happened to her heart and lungs in 1975 is her memory. This chapter has been concerned with how the person and her memory are able to tell us things that the person did not even know she knew.

3

The Personal History

The diagnostician whose questioning has brought to light what is happening to the patient's body has accomplished only part of the task. Three more areas must be investigated. The first concerns who the patient is, and how the kind of person the patient is, along with how he or she behaves, interacts with the pathophysiology to produce this specific illness. Next must come an attempt to discover whether other factors, environmental, familial, social, occupational, or personal habits, have played a part in making the patient sick. The final area concerns how the patient defines the problem; what has to be made right before the patient will consider the problem solved.

Throughout volume 1 I showed how a person is portrayed in every aspect of his or her spoken language. Listening to patients describe symptoms and respond to questions provides the opportunity to observe the paralanguage and hear the choice of words, syntax, premises, and logic by which speakers make themselves known to the skilled listener. However, patients have more explicit tales to tell, if only you ask—those elicited by their Past History, and Review of Systems. This chapter concerns these aspects of the history. More is involved here than an increased understanding of who the patient is. Sometimes illness occurs because twenty years earlier there was an exposure to asbestos, and in the intervening period the person smoked cigarettes heavily, or because of heavy alcohol consumption, or a hereditary susceptibility to coronary artery disease. Perhaps the patient has been taking estrogens for many years. Marital status, recent divorce or death in the family, the responsibility for many children, are examples of events or processes that may predispose to, aggravate, or even ameliorate illness. Patients may consider themselves to be so like a particular parent that they believe they are destined to follow that patient's footsteps even unto death. All this information, and much more, is contained within the parts of the history discussed in this chapter.

In an earlier chapter I discussed how the patient's interpretation of symptoms may make it difficult for the physician to discover what is really happening in the body. In the review of personal history, family history, and past illnesses, other factors may obscure the facts. First, the patient can be entirely unaware of the forces that have been acting on him or her over the years to produce illness. Consequently the patient may be impatient with the doctor's questions and therefore provide incomplete or incorrect information. Patients may be embarrassed about their backgrounds. Not realizing the influences of social class, place of birth, or the social or job status of parents on their own health, they may conceal or distort information. Sometimes illnesses that carry a social stigma may be concealed. These problems are compounded because doctors may share the patients' biases. They may believe that certain people do certain things just because they belong to particular groups. Jews are . . . Italians do . . . Blacks always . . . alcoholics often . . . ironworkers always. . . . When we physicians adopt stereotypes, we become blind to the uniqueness of the person sitting opposite us. Because of the blindness that bias introduces, we tend to disregard information about a particular patient that conflicts with the stereotype.

Today, however, people do not want to consider themselves biased. Certainly we do not want others to suspect that we might be bigoted. Because of the fear of being called a bigot (even by ourselves), physicians often disregard a vital and rich source of information—the knowledge that different groups behave differently. In fact the diet of people of Southern Italian background is likely to be very different from those whose people came from England. Young, college-educated, middle-class women in New York City dress differently from clerical workers with high school education. The former smoke less than the latter, tend to marry later, more often have children at an age when amniocentesis will be required, and (lest you have any questions about the applicability of this information to medicine) tend to question their physician's judgments more often, take fewer prescribed medications, tend to worry more about side effects of the drug than the dangers of the disease, and much, much more. If you do not know those facts, you are missing out on a tremendously important source of information about sick people. In similar fashion homosexual males are at risk for a host of infectious diseases, such as ameobiasis and pneumocystis carinii, which are not prevalent in the remainder of the population. The risks may vary with numbers of partners, types of partners, types of sexual contacts, and other factors not generally known to the heterosexual ("straight") world. Everyone knows that sexual behavior is one of those aspects of human activity that may be so shrouded in secrecy and misconceptions (let alone deceptions) that the truth is hard to find. The truth is difficult but not impossible to discover—and to physicians, vital. As the years go by, doctors come to learn an enormous amount about human behavior: about how people dress; about the motions required in the work of lawyers, draftsmen, hairdressers, and prostitutes; about diets and food fads; about fancies and fears; about the relations between parents and children, and what it means to lose a father, or mother, or a child; about sex and its pleasures and problems; about how people of a multitude of sizes, shapes, aptitudes, desires, backgrounds, and life-styles lead their lives and come, thereby, to risk illness or to recover. Doctors discover all of this wonderful information by watching, listening, and asking questions. Much of this knowledge is acquired in the course of taking the personal, family, and past history.

But how do we reconcile the need to avoid stereotypes while at the same time utilizing the knowledge about patients that comes from an understanding of variations in behavior based on cultural, social, or group differences? Stereotypes are damaging when we accept the information provided by the stereotype without making sure that the facts apply to *this* particular patient. A piece of information is just that and no more—it is not carved on a stone tablet. If a second fact contradicts the first, then more information must be sought to clarify the issue. A certain flexibility and willingness to change one's mind is essential to evaluating information from many sources. (It is easy to write the word "flexibility," but flexibility is one of the most difficult traits to acquire.)

In this part of the history a different principle of questioning applies, that of the unchanging question. In the previous chapter I encouraged changing from open-ended, to leading questions, to misleading questions, to yes-no questions, to fixed alternative questions ("Is it this or that?"); any question that would help clarify otherwise obscure answers was employed. But here the task is different. In this part of history taking one employs a relatively standard list of questions to find out as much as possible about people's habits, allergies, birthplace, parents, jobs, exercise, previous illness, sexual activity, and so on. Often the desired information is very personal, and some questions may appear to the patient to have no relevance to the illness situation. Further these aspects of the history require time and attention from someone who, because of illness or other reason, may be fatigued and inattentive. Conversely, these questions may give some patients license to tell all about things in which the doctor has no interest—occasionally at great length and in maddening detail.

Let me illustrate the direction I shall be taking. A new patient arrives on the hospital floor and goes to the nurse's desk where he makes a telephone call. He walks down to the sun-room, stopping to speak to the medication nurse (addressing her by her first name), and then finds fault with the television set and the lounge chairs. The charge nurse spends some time trying to find him and finally gets him back to his room. On his way he asks to make another telephone call at the nurse's station. I think there will be problems caring for this patient. Further I believe every experienced charge nurse will agree. What kind of difficulties? I do not know, but during his stay on that unit the staff will *know* he is there.

How can I make such a prediction without further knowledge of the patient, his disease, and the hospital? After all, what did he do. Everybody makes phone calls, first name usage is common nowadays, and hospital lounges frequently have poor chairs and less than perfect television sets. The predictive value of his behavior arises not because of his specific actions but because his behavior is so different from the usual new patient arriving for the first time on a hospital floor. Most patients go to their rooms and are quietly compliant with the few demands made on them. They usually wait to see what is expected. Our patient's acts are not unusual in themselves; only in their setting. Further hospital staff have seen many hundred patients arrive on their units, and though they may not be aware of what constitutes usual behavior, they are quick to spot differences.

The point of this story is that the doctor asking the social, past, and family history wants to be like that hospital unit—so much a constant that one patient's responses can be measured against the responses of others. This is a difficult task because many of us change (within limits) from day to day, making true constancy impossible. Much can be accomplished, however, if the goal of making oneself into a fixed measuring instrument is kept in mind. In the years when I designed questionnaires for use in a study of the health effects of air pollution, I learned what everyone learns in questionnaire research—if you change the wording of a question, you may change the response. One may think the question is basically the same—after all only a few words are different—but any change makes it a different question. Because of this all questionnaires must be extensively tested before they become suitable research instruments.

In history taking the doctor is the measuring instrument. Thus each physician should carefully choose which questions to employ, their exact wording, and the order in which they will be asked, and *then stick with the choice*. Not a word should be changed, not even the paralanguage should be altered, without good reason. Obviously this is a difficult goal, especially for beginners. But it can be done. More important, the concentration required in history taking and the attention to detail contribute to the development of history-taking skills. Remember the object: not only that the answer provides information but also that it can be compared to the answer of every patient asked the same question. Whenever I change a question or introduce a new one, I think about it and work on the wording for quite a while. Then I try the question out, making changes until it elicits the information I am after. Follow-up questions can be used to pursue particular facts, but the lead question should be unvarying.

Let us now go through an entire history, examining each question. The patient is a divorced woman in her early thirties. She came to see me because she had discovered a mass of lymph nodes in the left side of her neck. Four years earlier a histiocytic lymphoma was discovered in her right hip which was radiated. Eight months prior to the present visit, bony metastases were discovered, for which she received some kind of chemotherapy in Brazil. By her description it was unorthodox therapy. Follow-up X rays were said to have been negative. An enlarged left cervical node also appeared at that time but was apparently not considered important. About a month before this visit, a large mass of nodes appeared in her neck. She had seen several physicians before coming to me, and one of them was exploring the possibility of an infectious etiology because of her extensive travel in the Far East and Africa in the course of her work. She was referred to me by a clinical psychologist who had known her for a long time and who was very distressed at the management of her illness. He called to ask that I see her without delay. At this point in her history I have already obtained the story of her present illness. The only other information about her that is available to me is her name, address, next of kin, source of referral, and medical insurance, special billing instructions, occupation, and place of employment—all of which appear on a form from which my billing is generated. She has also completed a Review of Systems. I usually do not look at the Review of Systems until I have finished my oral history. Many years ago I employed an excellent, long questionnaire which allowed the patient to write the answers to all the questions I am about to ask. As complete as it was, it had, for me at least, a serious disadvantage. I found myself reading a form rather than looking at the patient, and when all was complete, the questionnaire had the information, not I.

As she answers, I make notes on something I call a "History Scratch Sheet." One side is blank, and there I make notes about the present illness. The other side has the abbreviated questions with room to write the answers. Since I dictate all my notes for typed transcription, this sheet remains a rough record of the initial visit. There are many other methods for keeping notes, some simpler, and others more complex. Over the years this method has evolved to suit my needs. The form also serves to remind me of the next question, although by now the order is etched in my memory.

The first questions relate to habits.

(In this transcript, no attempt has been made to duplicate the natural speech.)

EJC Do you smoke cigarettes?
Patient No.
EJC Do you take any medications regularly?
Patient Some vitamins and minerals. Do you want to know the list?
EJC Yes.

(In this instance I am interested because of the story of unorthodox therapies. In another patient I might merely note "vitamins," or even nothing, depending on time constraints and the importance in the particular case.)

Patient Three thousand milligrams of vitamin C. Some A and E, and a mineral compound thing and eight calcium tablets.
EJC How much A and E?
Patient I'm not sure.
EJC How much do you drink . . . alcohol.

(My question on alcohol intake may seem silly in view of the complexity that can attend a drinking history. But it is precisely this question that should make a crucial point about this aspect of history taking. Without using a long set of questions, how does one get an accurate picture of alcohol intake, especially in view of the denial characteristic of alcoholics? I have settled on this ultrasimple opener. Over many years, asked in precisely the same way each time, this question serves to elicit an answer that can be compared to thousands of other responses. People for whom alcohol is unimportant are quickly separated out. They respond without much pause: "Maybe I'll have a drink at a party," "I like a glass of wine with dinner sometimes," "I don't like it, although if I have to I'll have a drink ... like at a wedding or something," "On weekends, if we go somewhere, I'll have a few drinks," and other, similar answers. In the next category are regular drinkers: "I have a drink or two before dinner, and wine with my meals," "Pretty regular, but I don't think it's too much," "I like a cocktail," "A drink helps me unwind when I come home." Answers such as these suggest habituation, so I may pursue the matter, in order to see if alcohol dependancy is present and to determine amounts of alcohol. "When is the last day that you did *not* have a drink?" "What does a drink do for you?" "How does your drinking now compare to two years ago?" "How do you have your drinks (cocktails, highballs, straight)?" "Who mixes the drinks in your house?" "Give me an idea of how you have your drinks." Notice that the questions are open ended rather than of the yes-no type. This gives one an opportunity to hear what is not being said, as well as what is. If further questions reveal dependancy, I may ask questions to see how much tolerance has developed. One should remember that the point of questions about alcohol is to find out whether alcohol is a present or potential contributor to dysfunction, either through its contribution to other diseases, or through alcoholism, itself. Since denial is a classic feature of alcoholism, one cannot expect truthful answers to questions that the alcoholic considers to be traps. But if you slowly build the history, this usually allows you to discover that someone is having trouble with alcohol before the patient knows the information has been revealed. The questions themselves should tell the patient that the doctor is *not* making moral judgments; thus if further discussion ensues, the patient already knows that a sympathetic listener is present. Here, as in all history taking, the more one knows about a disease and about human behavior, the better the follow-up questions will be. All of this was introduced by the question:

EJC   How much do you drink ... alcohol?

Patient   None. I stopped drinking anything when I started getting my treatments in Brazil. I didn't want to interfere with my body anymore.

EJC   Are you allergic to anything?

Patient   No.

EJC   Are there any foods or drugs that make you sick, make you break out, give you pain in your abdomen, or that you avoid for any reason?

(That mouthful and the question before it are primarily meant to elicit drug reactions. While I am also interested in other allergic phenomena or food intolerance, I am doing my best to avoid causing a drug reaction. Obviously an allergist or any physician treating a patient with an allergic disease would ask more extensive questions.)

Patient   I don't think so.

(Usually at this point I would ask "Have you ever had X-ray *treatment*, or radiation *treatment*—not pictures, but treatment?" However, in this instance, since her original lesion had been radiated, we had already discussed radiation at length.)

EJC   What is your usual weight?
Patient   One hundred and fifty-five pounds. I'm a little heavier now; that happens a lot in the winter.

(If the patient does not volunteer that information, I ask if there has been any recent gain or loss. When weight control seems to be a problem, I add "What is the most you have ever weighed as an adult? When was that? What is the least you have weighed, and when was that?")

EJC   Where were you born?
Patient   In Duluth.
EJC   What kind of work do you do?
Patient   I'm a designer.

Where the potential for work-related illness or occupational exposure is revealed, follow-up questions are employed. In asking these questions, we learn about the working conditions of our patients. In one set of office hours I saw two patients who correctly attributed their symptoms to job stresses. The first was a sanitation worker and the next, the president of a huge corporation! This juxtaposition of two men with occupations at opposite ends of a scale of job responsibility, both of whom suffer from work-related stress, should make it clear that it is how the patient perceives the environment that determines what will be considered stressful, rather than some objective measure.

You should also be aware of how equipment is operated, what kind of chairs are used, and the social conditions of the workplace in addition to the potential for toxic or hazardous exposure. I dare say that every occupation has some problem that is unique to it. You should not hesitate to ask detailed questions about the patient's work. With few exceptions people love to talk about something in which they have expertise, and they are usually glad to teach their doctors. This section of a history should be considered to consist of introductory questions, where further questioning will follow when necessary.

EJC   How much education have you had?
Patient   College.

Did the patient drop out of college, not finish high school, go on for advanced degrees. Is the educational achievement beyond or behind job status, and if so, why?

EJC   Tell me about exercise.
Patient   I used to take dance classes, but since I travel so much, I hardly do anything any more.
EJC   Are you now or have you ever been married?
Patient   I'm divorced.
EJC   When was that?
Patient   In 1963.
EJC   How long were you married?
Patient   I think, ten years.
EJC   Was that your only marriage?
Patient   Yes.
EJC   Do you have any children?
Patient   Yes, two. I have two girls.
EJC   How old are they?

I then ask about the children's health, and I often ask the their names. I ask patients whether they have children even if they have always been single. Throughout the interview I am attempting to establish that I am not going to make moral judgments about what I am told. The evenness of tone and the automatic quality of the questions is meant to distinguish questioning from nosiness and to increase the accuracy of information, even with sensitive issues. If she had a spouse I would ask how long she had been married, was this the first marriage, and what was the spouse's age, occupation, and state of health.

The Family History

EJC   Are your parents alive?

Patient   My mother is.

EJC   How old was your father when he died?

Patient   Eighty-four.

EJC   What did he die of?

Patient   He had a stroke.

EJC   What kind of work had he done?

Patient   He was a lawyer.

EJC   Could you describe him briefly, please. What kind of a person he was, personality.

Patient   He was rigid, but sensitive and humorous, and probably a little crazy.

The usual family history elicits ages and causes of death in order to establish whether heritable diseases must be considered. As you can see, my questions go beyond those possibilities.

EJC   How old is your mother?

Patient   Seventy-one?

EJC   And her health . . .

Patient   OK, although she is always sick with something. Now she has arthritis.

EJC   Was your mother employed?

Patient   She had her own business, a lace store.

Questions in a history must change with changes in the social milieu. The women's movement had taken hold in the United States when these questions were framed. Asking women whether they or their mothers worked, might lead one into the trap of appearing to believe that housewives do not work. Caring for feminist patients who might already be wary of physicians was made more difficult if the patient believed me to be "hierarchical" or "sexist." What my personal beliefs might have been was not the issue. As the doctor learns who the patient is while taking the history, so the patient learns about the doctor.

EJC   Could you describe your mother briefly; what kind of a person she was.

Patient   She was a very difficult lady who was once spunky. I, personally, don't get along with her too well.

Asked about her father, one patient said "He was a baby"; and of her mother, "She was a baby, too." It takes little imagination or experience to guess that she considered herself the adult who "took care of everybody." Why does that matter? When such a patient becomes ill, she may be difficult to care for (as are many physicians), because of her image of herself as caretaker. On the other hand, such a patients may become extremely dependant when ill—after all it is finally her turn to be cared for. But whatever occurs, the doctor has been alerted to a possible source of difficulties. In this instance the fact is *not* that she considered herself the adult who took care of everybody. That is an interpretation. The *fact* is that she described both her father and mother as "babies," an unusual statement that awaits the addition of some other facts before it can become the basis for a conclusion or for action.

It is not contradictory to stress the importance to an individual of relationships with parents and at the same time to warn of the dangers of overinterpretation. It is presumptuous to believe that one can know "what kind of a person" someone is by asking a few questions. What the answers to these questions provide is information to be used in assessing the patient's reactions to illness. In asking these questions, I am attempting to discover what forces are acting on this patient that may be reflected in the patient's behavior in response to illness. I am also attempting to discover the meanings she may attach to her symptoms that arise from, for example, illness in the family, the patient's identification with a parent, the stresses stemming from social or economic ambition.

I do not believe it is an overstatement to say that the most important events in people's lives relate to the deaths of their parents, siblings, or their children. Thus one wants to know details. If the father died at a relatively young age, how old was the patient when he died? Was the patient orphaned? Did a near-age sibling die, and if so, when and how? Here, as always, try to make time for the answers, because the patient may be hesitant to reveal personal tragedy. But, gently and persistently, find out. It is a far different matter to suggest the diagnosis of depression to someone whose mother and two siblings committed suicide, than to one whose past is unclouded by tragedy. Because relationships to parents and siblings may be tinged with negative emotions such as anger or guilt, maintaining a judgment-free appearance while asking questions may be crucial to obtaining the information.

EJC   Do you have any siblings?
Patient   I have a sister who is younger than me.
EJC   Is she well?
Patient   Yeah.

("Is she well?" does not serve as well as "and her health . . ." Here, the open-ended question would have elicited more information.)

EJC   Is there anyone in your family, including aunts, uncles, and grandparents, who has had cancer?

I then go on to tuberculosis, heart disease, diabetes, and hypertension (referred to as "high blood pressure"). I have recently added stroke. Other physicians may add diseases in which they have particular interest. The next question is a net cast for things I may have missed.

EJC   Are there any other diseases that run in your family, or are there things about your family history that cause you concern about your health?
Patient   CANCER!

This is by far the most common answer, but patients may be concerned about anything from mental illness to glaucoma. Sometimes they are concerned about diseases or conditions that you never heard of, or even which will make you laugh. Try not to laugh, although an attempt to change the patient's beliefs might include some gentle teasing. If someone has been raised from childhood to middle age, however, believing that "all the Thomases die on holidays," the chances are that one visit to a new doctor is not going change that person's mind. Just try to remember that part of the history when you meet the patient in the emergency room with chest pain on Thanksgiving! If you do remember, it is very effective (if true) to be able to say, "This is one Thomas who is *not* going to die on Thanksgiving!"

Questions in the family history, as in all other sections, may be changed to suit regional beliefs, particular patients, and the changing times. In New York City, at least for the past decade, patients have accepted the questions about parental personality without difficulty. Occasionally, asked what a parent was like, a patient will supply physical characteristics. Other patients, not accustomed to American medicine, or used to impersonal care, will wonder what that has to do with medical matters. I answer that most people want to be treated like the persons they are, and knowing about their mothers and fathers helps in knowing about them. But most often (and, when I first tried the question, surprising to me), the person gives a description using only a few words. In those instances when I have known the parent (when the parent is also my patient, I say, for example, "While I know your father, it is what you think that interests me"), the descriptions are most often very accurate. Moreover one commonly comes away with a fair idea of which parent the patient identifies with.

In some sections of this country the history taker should be aware of regional diseases, occupations, family or organizational ties that might affect how illness develops, is recognized, or will be cared for. In small towns the doctor, in asking these questions, is making it clear that he or she is specially interested, or specially knowledgeable about the patient's family or other local issues. But in these cases physicians must be careful to let the patient do the talking. It is necessary that the patient be assured of privacy, and of being an individual, not merely somebody's son or daughter. This is crucial for children, adolescents, and young adults. Whether young persons come to see you with serious sexually related problems or illnesses, rather than going to a clinic or worse when they most need you, may depend on how well you have established your concern with them as individuals.

The Past Medical History

EJC   Have you ever had whooping cough?
Patient   I think so. Yes, when I was little.

I then go on to scarlet fever, asthma, and rheumatic fever. I ask about pertussis because it occasionally has bearing on later diseases of the chest. But why ask about scarlet fever (or diphtheria) or a host of other diseases that usually have no bearing on the adult disease pattern? Scarlet fever may be relevant only if renal disease shows up in the remainder of the history or workup. Too many questions are asked because of stale custom. They should be replaced by questions that are still pertinent to medical practice.

It is surprising how many patients have a history of asthma which they remember during their first visit but which they have forgotten when they come into the office wheezing at some later date. I have not discussed how one prevents information acquired at this time from dropping into obscurity. In the future computer retrieval will make such accidents of memory less common, but until then one must devise one's own method or use one of the preprinted versions of records that feature ways of keeping key information in plain sight. I mark the front of the chart folder "History of Rheumatic Fever," or "Allergic to . . .") In addition I now paste a copy of this section of the history on the inside front cover of the chart so that it is always in view.

EJC   Have you ever been told of high blood pressure or hypertension?
Patient   No.
EJC   Have you ever been told of heart disease or a heart murmur?
Patient   No.
EJC   Have you ever been told of diabetes?
Patient   No.

Notice that I do not ask, "Have you ever been told of high blood pressure, heart disease, or diabetes?" A yes or no answer would not tell you to which disease the answer applied. Also note the distinction between hypertension and high blood pressure, which are seen as different by some patients. In other patient groups one might include the words "or sugar in your blood or urine" in the question about diabetes. Again, regional or population differences dictate different questions, or follow-up questions after the standard form.

EJC   Have you ever been operated on?
Patient   I had my appendix out.
EJC   How old were you?
Patient   Just a kid.
EJC   Any other operations?
Patient   My tonsils, and I had a boil opened on my back once. Then, of course, I had the biopsy on my right hip.

Unless the relevance to the patient's current illness is obvious, it may be tiresome to ask these questions. I have never seen the necessity of pinning down the the exact age at which the patient's pilonidal sinus was repaired, nose straightened, or ears bobbed. In hospital records, such information may later be required for research. I am more concerned with patients' major operations and their outcome, and with their reactions to the experience of surgery, than I am with a detailed exactitude about the unimportant. On the other hand, we all learn to take histories before we know what is important, so that being overly compulsive early on may be the lesser evil.

I go on to ask, "Have you ever had any other hospitalizations for physical or mental problems," and to questions about accidents, fractures, or periods of unconsciousness. Men are asked whether they have been in the armed services, and if so, whether they were in combat. For women, I ask about the onset of menses, or time of menopause, depending on their age.

EJC   How many pregnancies have you had?
Patient   Uh—three.

(Remember, we know that she has two children.)

EJC   Did you have any difficulty with the birth of your children?

If the patient had a history of hypertension, I would have asked whether she was hypertensive during her pregnancy.

Patient   No, it took hardly any time. They were wonderful.
EJC   And the other pregnancy?
Patient   I—uh—uh—it was an abortion.
EJC   When was that?
Patient   In 1967.
EJC   Things were not so easy then.
Patient   It certainly was not.

Abortion was legalized in New York State in 1970. Abortions in prior years were illegal, exceedingly common and often very traumatic, both physically and psychologically. Often, considerable guilt follows abortions, even when they are legal. Every physician must remember that people *never* forget traumatic events. The memory may be repressed, but it is *never* forgotten. The force of that fact will be evident to you the first time you meet a patient still making amends to the world or her family because she institutionalized a retarded child forty years earlier! More traumatic events are found in relation to sex and pregnancy, I would guess, than in any other aspect of the human condition except perhaps relations with family. Often these events are deep, deep secrets. It is for these reasons that questions here must be extremely neutral. I ask, "How many pregnancies have you had?" of adolescents and the never married as well as other women. In addition to increasing the chance that important illness events will come to light, it provides another opportunity to provide a morally neutral ear, especially in case the patient should want to unburden herself.

EJC What kind of contraception do you use?
Patient None, since my menopause.

I would usually not have asked her, but I forgot that she was post menopausal. When oral contraceptives are employed, I question the duration and type. When they are currently not used, questions about past use follow.

EJC How often do you have sex?
Patient It varies. Now, maybe two times a week. It depends on who I'm with.

Sexual histories are notoriously difficult to obtain. This simple question, asked matter of factly, has served me best. Note that I do not ask, "How often do you have intercourse?" I am trying to get an answer from everyone. The answers vary widely, but I think two generalities can be made. The response of the majority of patients suggests no sexual problems, whether that is actually the case or not. In some patients, however, important difficulties are revealed, or the patient is given the opportunity to ask questions that would otherwise have remained hidden. Whatever their own sexual problems may be, doctors are considered by layperson to be knowledgeable about sex. Nevertheless, the subject is embarrassing. Thus when the physician provides an opportunity, or makes it clear that he or she can offer a nonjudgmental ear, the patient may bring up issues or symptoms otherwise not disclosed. While in my office for her second visit (a routine physical examination), a married women attorney asked, "Are you supposed to have pleasure when you have sex?" "Yes," I said, "why do you ask?" She told me about her anesthetic, anorgasmic sexual relationship with her husband. It had been a problem for a number of years. "Why didn't you talk to me about this the last time you were here?" "Well," she said, "you had that other doctor visiting, and I did not want to talk about it." "Why didn't you call, or come in again and talk to me about it?" "I figured it would wait until my physical this year." It is sad to contemplate the amount of unhappiness that remains buried because patients feel themselves unable to speak to their physicians. Every step of your history should make it clear that you are a doctor who listens—and understands. My final question is "Have you ever had syphilis or gonorrhea?" I suppose that I ought to include amebiasis, herpes simplex, hepatitis, cytomegalovirus, and pneumocystis carinii, but I just have not gotten to that point quite yet, fearing that the list will grow endless. Yet the prevalence of these diseases has risen dramatically in recent years among homosexual men, causing them great anxiety. Homosexual men may avoid asking their doctors questions about these diseases, out of fear of the physician's reaction to homosexuality. At least at this time it is difficult to include a question in the sexual history that asks whether the patient is heterosexual, homosexual, or any combination thereof. To the homosexual patient such a question may be heard as an accusation, no matter how it is asked. The heterosexual patient may wonder whether the doctor is asking the question because he or she knows something about the patient that the patient does not know.

EJC   Are there any other diseases, illnesses, health conditions, or health problems, past or present, that you wish to tell me about?
Patient   I was once told that I had an ulcer, but that was a long time ago.

I clarified that issue and then concluded with the following mildly apologetic acknowledgement of the extent of questioning.

EJC   Between all these questions and that [Review of Systems] form, have I asked enough questions?
Patient   You certainly have.

The Quick and Dirty History

Many times, particularly in emergencies, one cannot take a long history. Yet it is essential that one discover whether the patient has other disease, allergies, or is taking any medication. The following past history "short form," which can be asked as a physical examination is being done, or even as a dressing is put on a bleeding wound, will get the information necessary for emergencies:

Do you take any medication regularly, even vitamins?

Do you have any allergies?

Are there any foods or drugs that make you sick, make you break out, give you pain in your abdomen, or which you avoid for any reason?

Have you ever been operated on?

Have you ever been in a hospital?

Do you have any other illnesses at this time?

Have you ever been seriously ill?

When is the last time you saw a doctor, for anything?

That brief series of questions will provide enough information about the past history to know whether the person whose immediate problem you are treating has had previous serious disease. Notice the last question: "When is the last time you saw a doctor, for anything?" There seems to be a bias among emergency room staff that makes them believe that their patients are well (or faking) until proved otherwise. You yourself may believe the patient you are seeing with, say, a headache, really is not seriously ill. But if, in responding to the last question, the patient says "I never go to doctors. I haven't seen one in five years," be cautious. Part of your diagnosis must include the answer to the question, "Why does a patient who never goes to doctors come into an emergency room for a headache? Sometimes it is because the headache is a symptom of overwhelming stress, and sometimes it is early meningitis. But the probability is nontrivial that something is wrong with that patient.

No matter how terrible the emergency, no matter how self-evident the problem, if the patient can speak, some history must be obtained. Again and again, serious errors are committed because the patient was not asked whether the present illness had occurred previously, or was not asked even those few questions listed here. I learned that lesson when, as a medical student, I was working in the emergency room of a community hospital. The noise of a group of people hurrying to the emergency room could be heard as soon as they had entered the front door of the hospital. Soon they appeared, all members of one family, carrying in their arms a young man whom they knew had appendicitis and they put on the stretcher. Shortly before going to the operating room the resident asked me something about the young man's past history. I did not know the answer. I had to admit that I had been so carried away by the family's diagnosis and their panic that I had neither taken his history nor examined the patient; neither had the surgical resident. After examination, the young man was not operated on; he did not have appendicitis!

The Review of Systems
Asking the complete Review of Systems can occupy more time than the remainder of the history. In my office patients fill in a form that I have used for almost twenty years. The questions were derived from other Reviews of Systems and from the questionnaire employed in epidemiological studies in which I participated for many years. Good questionnaires are hard to find for the reason that good questions are difficult to formulate, questions that ask precisely what you want to ask and no more or less. For the beginner the Review of Systems is both a burden and an opportunity. It is a burden because it takes so long and because of the occasional patient who has a symptom in each and every organ system! The opportunity arises because the student begins to *hear* what symptom reports sound like. Early in medical training physicians learn to see what diseases look like, both in the patient and in the laboratory. Acute gouty arthritis is the archetype of a hot joint. Ascites looks like ascites, and an enlarged nodular liver feels like nothing else. The murmurs of the heart and the noises in the chest become familiar friends through the earpieces of the stethoscope. But in taking histories, especially in listening to a Review of Systems, the verbal expression of organ function and dysfunction is made known. Unfortunately most doctors pay more attention to what dysfunctional organs look like than what the patient's reports of those dysfunctions sound like, unfortunate because the patient's report is usually how the illness is first presented. Therefore as burdensome as a full Review of Systems may be, it is something that must be learned. On the occasion when I must do the Review of Systems, I usually ask the questions while I am doing my physical examination.

The Patient's Definition of the Problem

At this point in the history a physician should have formed a working hypothesis (or hypotheses) concerning what is wrong with the patient, who the patient is, and the contribution made to the illness by the patient's life-style, occupation, environment, past history, and family history. There are, however, at least three other pieces of information that are particularly important: the patient's definition of the problem; the solution(s) which, from the patient's point of view, would be adequate; and the methods employed by the patient to cope with the illness. I inquire about these areas when I am taking the history of the illness, but they can be investigated at the end of the entire history as well. It is well worth reading Dr. Mark N. Ozer's detailed discussion of these issues in his book, *Solving Learning and Behavior Problems of Children.*

Physicians generally define their patients' problems in terms of disease; it is not surprising, then, that they look for solutions in the same terms and consider the problem resolved when the disease has been treated. For pneumonia or other acute illness, such solutions are often superb and sufficient. But consider the patient who, despite long-term or recurrent symptoms, is *not* found to have a disease and yet continues to see doctor after doctor. The fact that a patient persists in seeking medical attention even after being told that no disease is present suggests that the solution to the patient's problem has not yet been found. The second physician in the patient's odyssey should have been aware that the conclusion, "There is nothing wrong with you," does not satisfy the patient. In such instances a clear understanding of what concerns the patient—what the patient thinks the problem is—might lead to a solution. As this next example shows, patients with persistent bowel symptoms and repeatedly negative X-ray or endoscopic studies do not need yet another barium enema.

What concerns you most about your symptoms?

Well, I'm not worried that I have cancer; I already know I don't have that. But, you know, I never know when my stomach is going to make trouble for me. Like if I'm in the theater and I suddenly have to go. I mean I HAVE TO GO. And If I'm in the middle of the aisle, its a nightmare.

Um-hmm

So if I only knew how to CONTROL it a little, I would feel better.

Well, I think I can show you how to control it for special occasions, certainly.

That would be great. And the other thing I think is, what if it keeps getting worse and worse as I get older, won't it injure my body or something?

It won't, simple as that. Mostly functional bowel symptoms—what you have been told is spastic colitis—get better when you are older; or at least people learn how to live alongside of it. And they don't hurt the body or lead to cancer, or any of it . . .

In the remainder of the conversation, the patient was instructed to use an occasional small dose of codeine or paregoric for those special occasions when diarrhea would be a disaster. In addition the doctor spent time explaining in detail how this patient's bowel differed from a normally functioning bowel. In this way the discussion was moved from disease terms to terms more relevant to the patient's problem, bowel *dysfunction.* Currently we would suggest some form of stool bulk as a management strategy and explore whether or not diarrhea followed ingestion of milk products. By the time such a patient is seen, however, he may have been tested already for lactose intolerance and numerous other rare causes of diarrhea—a diagnostic safari that takes much longer than the ten-minute conversation necessary to meet *his* concerns and solve the problem *in his terms*.

Taking the time to inquire about patients' concerns ultimately leads to knowledge about managing the myriad ills for which there are no adequate diagnoses. If one keeps seeing things about which one knows nothing, one either finds the necessary information or dismisses what one sees. The former is the better route. For example, it does not help patients with headaches and other musculoskeletal problems to tell them their problem is "tension" or "nerves." If they were not too polite, they might point out that the right side of the head is as subject to "nerves" as the left, so why does only the left hurt? Perhaps it hurts because the briefcase is carried in that hand and the left shoulder is held higher than the right. Over the years the muscles on that side have shortened and are tighter than are the muscles on the right. When "tension" is associated with tightening of the muscles, those on the left side of the neck and shoulder can more easily be made to hurt because they are already tight. If that is correct, the pain can be made to go away, despite continued stress. Or perhaps the pain is due to some other reason that stubborn persistence in questioning or examination will reveal. A multitude of such examples exist.

Patients with definite disease diagnoses may also have concerns far removed from those of their doctor. Patients with malignancies may be less worried about dying than about being burdensome to their families. They may be aware that they will die but would like to try and reach the day of a child's graduation. Patients with clinically important heart disease may be more concerned with their osteo-arthritic joints, a matter of indifference to a physician concentrating on improving left ventricular function. All of these issues can be brought to light by asking about the patient's concerns. Dr. Ozer believes one should press further than one question; he suggests that the physician ask three times, "And what else concerns you?" After having elicited several concerns, the doctor asks, "And which of these is your *greatest* concern?" In this way the patient is given a degree of choice in treatment when the illness would seem to preclude choices. Any method that gives patients a sense of control can be very useful.

In chronic illnesses, or where symptoms have been long present, patients often work out methods for dealing with their disabilities. I ask, "Are there any ways you have found useful in dealing with . . .," or "Have there been times when you have been able to control your (symptoms or illness), and how did you do it?" Such questions will not only give you a further idea of the impact of the illness or symptoms on the patient but will tell you about the patient's resourcefulness, and may reveal avenues that should be pursued in treatment. Such questions are also helpful in the presence of patients who feel depressed or helpless in the face of their illness; by teasing out the patient's coping mechanisms, the doctor can demonstrate that the patient *does* have the ability to cope with problems that he or she sees as insoluble. The doctor can also help clarify the resources used by the patient in coping.

A related but distinct set of questions enables the doctor to find out what the patient would consider an adequate solution to the illness. Once again, if the care of infectious diseases or acute trauma is used as a model for medical care, such questions may have little applicability. But present-day patients more often have chronic diseases or illness states that cannot be cured or symptoms that can only partially be relieved. Only the patient can know what would give the greatest relief or return to function, and, when possible, this is where the attention should be directed. Sometimes when patients present several different illnesses at the same time, I use a hypothetical question to find out what is most important to them. "If your skin was clear, but your back continued to hurt, would that be satisfactory?" "Oh yes, I'd do anything to have that itch go away. I can put up with my back; I have for years."

At this point you and your patient can go to the examining room with the assurance that the most crucial part of the visit has been accomplished. You should, by now, have solid ideas about what your goals are for this particular patient and what actions will be required in the near future to meet those goals. Your patients will have the sense that you understand the problem and that you understand and care about them.

Reference

Ozer, M. N. *Solving Learning and Behavior Problems of Children.* San Francisco: Jossey-Bass, 1980. See especially ch. 4, 5, and 6.

4

Personal Meaning

The story of an illness—the patient's history—has two protagonists, who are intertwined: the body and the person. By careful questioning, it is possible to separate out the facts that speak of disturbed bodily functioning—the pathophysiology that gives you the diagnosis. To do this, the facts about the body's dysfunction must be separated from the meanings that the patient has attached to them. Skillful physicians have been doing this for ages. All too often, however, the personal meanings are then discarded. With them goes the doctor's opportunity to know who the patient is.

In taking the social, family, and past history, the physician begins to discover the social and environmental forces acting on the patient. These forces may influence how the illness was acquired; they may also influence its course, treatment, and outcome. These aspects of the person are, paradoxically, somewhat impersonal, since they have primarily to do with factors (such as cigarette smoking or allergies) that act on someone from the outside. There is another aspect of the person—personal meaning—that is vitally important to under-standing how a person and a disease interact to produce a particular illness. In volume 1 I discussed how meanings are assigned, how they influence human behavior, and how complex the phenomenon of meaning is. Now I would like to show how meanings manifest them-selves specifically in a patient's story and how the physician can uncover them.

Although specific questions must be asked to obtain the social history, clues to important personal meanings drop in your lap whether or not you want them. The trick is to pursue them with further questioning. The process is much like using questions to find out what bodily function has been disturbed. There, your knowledge of basic clinical science—anatomy and physiology—guided your questions. Here, your knowledge of human behavior must guide you.

Unfortunately most physicians know more about how the liver works than about ordinary human behavior. We teach much about livers during medical training but little about people. In addition few physicians realize the vital importance to clinicians of knowledge about day-to-day human behavior. We do have a referent, however, that can serve us very well as we explore the details of a patient's behavior—ourselves. We should constantly be asking ourselves how we would have thought, felt, reacted, or acted if such an event had happened to us. Would we have been frightened, angry, run away, gone to a doctor, taken the medicine? If the patient responded to the stimulus differently than you believe you would have responded, then the reasons must be sought. However, in order to use yourself effectively as a reference against which to understand the patient's actions, you must *not* make a value judgment about what the patient said or did. The words, "dumb," "stupid," "ridiculous," or even "smart," "clever," "intelligent," must *not* be applied to what you hear (although avoiding such judgments requires discipline). When such judgments do intrude, they must be forced out of thought because they interfere with understanding the patient. These value judgments are akin to substituting an interpretation for an observation and thus losing sight of the observation. When a patient tells you something that makes the word "stupid" rush to your mind, you must stop and remember that there must have been a reason for what the patient did, or the patient would not have done it. Further you want to know that reason, the underlying premise or belief on which the patient acted, because it will probably raise its head again in the future.

Let me make the point more clearly with an example. We met this woman in chapter 2, as I showed how careful questioning could bring out the story of thrombophlebitis followed by pulmonary emboli.

I was feeling quite—
Right. Next question.
— miserable.
Had you seen a doctor yet with all these pains?
No, I kept putting it off.
And then what happened?
Then I went to see a doctor.
Who?
He— well, really, he (Dr.——) wasn't an MD. He was an osteopath.
Mm-hm.

Do you know of him? No? Very good. He was more of a nutritional—well, he believes in vitamins and—

What did he say when he examined you. Did he examine you?

Well, it just so happened that he was away. He had a couple of other doctors who were MD's in his office.

Mm-hm.

He took a variety of— he has a very large lab in his office, so he took many blood tests. The sed rate the first time I was there was thirty-eight.

When was the first time you were there?

I think it was the 11th.

OK. What did he say was the matter with you?

Well, they also took some X rays of this and the doctor who was there thought there was a break.

Mm-hm.

Well, it turned out not to be broken. It turned out to be the old break. And I took some X rays over at Deever Hospital, a few days, or a week later, and that didn't show a break. It showed an old situation. Anyway, the lungs were inflamed, so they said. I had a pneumonitis or something like that. The sed rate was thirty-eight, and he gave me—I guess a Penicillin, or an antibiotic shot, and some Declomycin which I had been taking since then, twice a day—

You saw him on the 11th? When were you first told of the pneumonitis?

Well, it was sometime during that week, I think they were puzzled.

The patient's pulmonary emboli could easily have killed her. The fact that she received poor medical care does not seem to have been due to chance alone. She sought out the physicians and stayed with them. Was she unsophisticated or unknowledgeable? Apparently not. She knew her medications and remembered laboratory results. Whether because of denial, or other reasons, her behavior, interacting with a dangerous disease, might have cost her life. If this patient had died of her next embolus, one could not have considered it "psychosomatic," in the conventional sense of the word. People would have said "Rhoda was so dumb. She should have known better." Not only should she have known better, but, as we shall discover, she did know better. She knew what was wrong with her. From the information at hand, we do not know why she behaved in this manner, but we are given reason to be alert to similar behavior in the future.

This next example is from the young woman with ulcerative colitis that presented as proctitis (chapter 2). I have included a fragment from the history of the illness as well as the pertinent segment from the family history.

I think I had hemorrhoids.

Mm-hm. When was that?

But, about two months ago.

Mm-hm.

But and, it was— um. But then they went away. I didn't do anything for them.

Did you have bleeding back then, a few months ago?

No.

I see. And how did you know at that time that you had hemorrhoids?

I thought they were. Because I felt something there.

Did you feel something? A lump or something?

Yeah. Like, two little lumps.

(Later)

Are your parents alive?

Yes.

How old is your dad?

He's sixty-seven. Is he in good health? Not right now. As a matter of fact, he— this isn't something— he-he had the same thing recently. But his, I think, is colitis. And he went to Dr.——, one of the specialists.

And they looked up inside of him—

Mm-hm. And it might be colitis. But, I'm sure that has nothing to do with this.

Tell me a little about his—when did it happen?

I didn't even think of this until I asked my mother what was wrong with my father. And then I noticed I had bleeding, too. But I didn't think about it, and I just— and said it just couldn't possibly have anything—

I understand, too, but I'm trying—

It was just a coincidence—

I'm still interested in the time sequence. When did you know your father had colitis?

I'd say—last week?

AFTER the bleeding.

Yeah.

After the bleeding.

Then I thought, "Maybe I'd better see a doctor, too! Even though it might be totally unrelated." Which I'm sure it is totally unrelated.

Yes, I understand that. Of course, significance can go both ways, right? I mean, you might have just dismissed it as your hemorrhoid otherwise? Would you have, do you think?

I still do. Except I didn't understand why— when I had the baby, I had hemorrhoids. And again back two months ago I had hemorrhoids, but I never had bleeding.

Mm-hm.

And then when it was persistent for eight days, or like that, I wouldn't let it go.

Was it her rectal bleeding that brought Alyce to the doctor? Or was it her rectal bleeding *and* her father's "colitis?" People take action because of the meaning of events. Alyce's rectal bleeding acquired a new meaning after she heard about her father, and because of this she went to a doctor. I have no doubt, considering the eventual severity of her disease, that she would ultimately have sought medical care. She did what she did when she did it, however, because of a new meaning. Later in the chapter we will examine people's connection to their parents in more detail.

In chapter 2 a young woman was presented who had vomiting, bloating, and constipation for which she took many enemas. Here she is again.

Now, so, August, then it wasn't there in September and then in October it was there, but it wasn't bad like it is this week.

Right. I didn't bloat up. That's what really bothered me.

Now, I'm on vomiting now, see. I'm on vomiting. You mean, if you didn't bloat up now, you wouldn't mind vomiting.

—Not that much. I would— yeah, I guess I would actually I would.

Mm-hm. But I mean, if it was just vomiting, would you be here?

. . . . . Probably not.

You may wonder whether she is telling the truth. This next segment comes from a visit six months later.

A lot of things I have realized. But there's ONE THING—

Mm-hm.

—that I STILL don't have control over.

What's that?

Food.

Hm?

Eating. Now I realize that the reason that I had all that throwing up thing—

Mm-hm.

I cannot believe that I never— I mean, I guess it was too much of a thing for me to have to realize that over at the dance school, you know, they're all— I mean, I might be stretching this out a bit, but I thought, because I'm so used to having people say to me, "Oh, you're so good, you're so good—you've been doing so well!" And they didn't really say anything to me, and they never really said I was BAD, but they never said, "You're doing good work, except one of the teachers in the Modern department. And then when I switched out of the Ballet department into the Modern department, and it was like the second day, or something, that I was there, and she goes, "Well, you MUST lose weight. Five pounds minimum."

Yes, go on, "You must lose weight, five pounds minimum—" So?

And she didn't say how. So—

Do you think that's why you were vomiting?

I mean, I'm sure that's a— I mean, I KNOW the reason I was throwing up was— it's ridiculous, I mean—

What was the reason?

I wanted to be skinny. I still do.

To physicians, vomiting is usually an important symptom. Without knowledge of the world, many doctors are unaware how many young women voluntarily vomit in order to control their weight. Ballet dancers are barely rational about maintaining weight control. Yet, in the ballet world, there is nothing irrational about vomiting to keep down weight: ultra thinness is frequently essential to remaining in the ballet company. If a medical student vomited daily, it would have a very different meaning—a second-year student would probably interpret the emesis as the first symptom of an inevitably fatal disease!

These next two patients illustrate the fact that it is not the weight change that sends patients to physicians but rather the meaning of the change in weight.

The next example, a patient who had metastatic carcinoma to the liver from the large bowel, was presented in volume 1. She had a routine physical in her employer's medical department in November and was found to have lost weight and to be anemic. She was advised to have gastrointestinal X rays, but she put them off until . . .

Because I'm on a diet. Because, you know, sometimes I go on a diet to lose weight because as I say, I was heavy. I'm still heavy, but I was losing weight and I was— to me, I was glad I was losin' it, because I could stand it. So I started to think to myself, "But why am I losing it?" Because—

When did you start to think that to yourself?

About the end of December. And when the doctor told me I had lost, he said, "You know you lost twenty pounds." And he said—

OK, but the physical was at the end of November.

Right.

Mm-hm.

But these things— it was GRADUAL. It didn't, like—

It was nothing sudden.

It was nothing sudden. It was all a gradual thing. In a week's time, I realized that—"My dresses ARE getting a little loose. I wonder why I'm losing weight?" And then I realized, "It's my appetite. I'm not eating like I normally eat." Because I usually eat a big meal, then I have dessert, and, you know— This way I'm eat— I find that I've been eating small meals and no desserts. So this way I—

Did that please you?

Yes! I was losing weight and that was what I wanted to do.

Mm-hm.

So, anyhow, I find that I would eat like meat, vegetables, and potato— yeah, and then I would be filled. And I had no desire for dessert.

Mm-hm.

And I said to myself, "This is not me." Because I always love desserts.

Now, when did you first realize this wasn't you?

—I would say it was around the Christmas holidays.

Were you concerned about it at all?

No. I was delighted. Because I had no pain, and I had no symptoms of anything. You know, I felt all right, it was just that my appetite was shot.

Mm-hm.

The patient came to medical attention in November, but did not act until late in December when she acquired a "symptom." Weight loss changed from a desirable occurrence to a symptom demanding attention only when she stopped eating desserts. At that point she stopped being herself. Within physicians' frames of reference involuntary weight loss is *always* a symptom because of its possible meanings, but to laypersons a change in weight may have very different meanings.

In this next example a gain in weight prompted the patient's concern. She is describing a visit to her surgeon many months after colon resection for carcinoma of the bowel.

And, when they weighed me at Dr.———'s, they weren't even going to weigh me. Dr.———felt my stomach, and he said, "Great." And I said, "How do I look." He said, "Terrific." And when they weighed me, I'd gained about five pounds.
Mm-hm.
And my appetite is very, very small since my surgery. And I'm delighted. And I said something to the nurse, and she didn't say anything. Meanwhile, I asked—
Why did that worry you?
Because why should I gain weight when I'm eating so little?
Mm-hm.
And moving my bowels. So I neglected to tell you that I have terrible trouble with my back. For years and years.

The patient was perfectly happy to have a small appetite and no difficulty maintaining her weight. What made her focus on the back pain as a possible sign of recurrence (although it had been present for many years) was unexplained weight gain to which she attached dire meanings. Each of us is as much the collectivity of our meanings as our bodies are configurations of organs, joints, and muscles. Thus, when you hear disturbances of meaning, you are hearing disturbances of a person. Personal meanings can be very enduring, particularly when they are linked to personality characteristics. This next series of examples are from the patient who did not have a symptom until she stopped eating desserts. Her bowel was resected, but metastatic disease ultimately caused her death. On examination, she had a huge liver. The first segment is from the initial history. She describes something about herself in passing that is very important to her.

And what did the pain feel like?
The pain was sharp and constant. It would come and go. And I noticed especially that it would go when I sat down.
Mm-hm.
When I stand or walk, I would get it.
Mm-hm.
So I continue to work. I never missed a day's work.

"I never missed a day's work"—a very characteristic statement for some people. These are the people who say, "I've been at this place for twenty-five years, and I've never missed a day's work!" They come in sick and give everybody else their colds, but they do not miss work. Incidentally, why does her pain go away when she sits down? Because she is resting her liver on her pelvic brim. So she tells us two important things in one brief utterance—how big her liver is, and her relationship to work.

The next example takes place just postoperatively.

Shook ye up, getting that blood, did it?
No.
Or being tied down on that bed, Dr. Jaros...
No. The only thing, they upset me. They hung those two big bottles six feet in the air. An' I said, please gimme a pole, so, ye know, if—
Right.
—I needa go to the bathroom, I wanna see the nurse or somethin'. I said, by puttin' those two things on the pole I'm an invalid. I'm to the bed. I can't go anywhere. So I got a little upset without the pole... but anyhow they brought me the pole, an' I was fine.
Good.

Five days postoperatively, after being told that she will die of her disease:

Now could I go to work an... an... ah... do ah.
Return to work? My guess is you won't return to work.
I will.
Will not!
Oh.
But funny things can happen.
An... n... I can't go to work at all?
I don' know that... I told you I don't think you will.
I... I'd like to try an' go an' see what happens. An' fight it if I could.
Fine! I'm on your side, then.
'Cause I'm really... I'm really a fighter, an' I really would like to get back to work.

Later in the same conversation

I'll probably... I'll probably just go down hill all the way, huh.
I don' know what that means, do you? Whadya mean by that?

With the liver.

What does that mean, "down hill all the way"?

Well, I'm gonna try an' go to work, which I think I can do in the begining. Then, I'll see how I make out ... if I can do it. If I can't, I'll have to resign. Stay home. If I stay home, then ... I'm not good at home. I'm much better at the office. I got more time to think at home.

On the following day (and, perhaps surprising to you, less depressed):

What's the matter, lady, are you sittin' there discouraged?

Pardon me.

Are you sittin' there discouraged, are you?

No, not really. . . . I made up my mind what I'm gonna do and that's what I'm gonna do.

Okay. It just doesn't happen by itself, you know.

My sons know me. They know I'm a fighter, so they ...

Uhm hm. Well, fightin' is not a theoretical matter, madam. Fighten' means that, that you move yourself, ye know. Ye know, ye see when you decided that you're just going to stay in bed, I'm all for that. . . . But when you decided that you want to get up and go ...

Oh, I'm determined to get outta here.

A few days later:

Well, also, another accomplishment that I made today.

Yup, Uh hm

Which is the first time I ever did. Which is very difficult for me. I walked down to the second clock! Ye see, we go by clocks.

Did you really? All the way that ...

All the way down to the second clock! Not the first. My family walked me to the first the other night, but I walked to the second!

In this series we saw how a side remark of the patient, uttered during her initial history, described her relationship to work and characterized her behavior in that respect throughout the illness. Note also how "tying her down to the bed" made her feel like an invalid. Another patient, for whom events had different meanings, might not be similarly distressed by a fixed IV pole. The point to note is that, for this woman, a blood transfusion acquires at least part of its negative meanings not from the surgery alone but from the enforced invalidism dictated by an IV setup.

Many of the aspects of person I am now discussing present themselves all the time. You do not have to ask, merely listen. The same is true of the patient's body: it is constantly telling you things; all that is required is your attention. How someone walks, stands, smiles, the color of the skin, circles under eyes, appearance of hair and clothing, all provide information. When we pay attention, we do not tell ourselves that we are making a diagnosis, we merely say to our friends, patients, or colleagues: "Are you all right? You look awfully tired," or "Have you lost weight?" or "Did you hurt your back? The way you're walking ..." Just as our knowledge about matters of the body should be a basis for our diagnostic thinking; knowledge of the person should equally play a part.

These next examples are of people telling, in ordinary conversation, who they are. None of these utterances are in response to questions, the words just appeared in the course of interaction. The first man is a real estate broker in his seventies.

Look, I'm not the kind of a fellow who wants to hear you pat me on the back. Because, while I'm an egotist, and recognize a lot of things, I have a lot of faults. Because I'm an uneducated person. I use the word, "uneducated" because maybe there are certain things that I lack. At best I'm a high school graduate. That's all. I've gone very far in the field of finance and economics with men who hold high degrees, I sit around with men who account in this world, and they recognize my little talent that I have. And when I say "recognize," they count on me to go and do things.

Knowing who he considered himself to be figures highly in this man's care, as is discussed in greater detail in volume 1 (see also my discussion of the psychological reactions to physical illness in another text, (*Psychiatry in General Medical Practice*, Usdin and Lewis, eds., 1979). Here is another example.

So my husband would not think of going into the hospital in Alabama. We are Bostonians, originally, so it was either Boston or New York. And my son-in-law is a physician, and my daughter, E——, is a little Ph.D. psychologist. So we called them, and they said, "Come here, instead of Boston."

In this little fragment of ordinary conversation, we are made aware of how this woman feels about Boston in relation to Alabama, and even to New York. In addition we might suspect her low opinion of her daughter, or of female Ph.D.'s in general, at least in relation to male physicians.

This is a fragment from an earlier example.

I made up my mind what I'm going to do and that's what I'm going to do.
OK, it just doesn't happen by itself.
My sons know me, they know I'm a fighter, so—
Mm-hm.

What we inadvertantly tell others about ourselves is not always what we would like them to know about us. Next is a woman telling two contradictory things about herself: what she would like to believe, and what is in fact the case.

Have you ever had any operations?
Oh! I had plastic surgery six months ago. My first operation. I had a little minilift!
And are you pleased with it?
Very pleased. It was my first operation. . . . I'm pleased except for one thing, which you'll see. It's the doctor's fault and he admits it. I've got two giant hematomas, and he did beautiful surgery. I had a revulsion against the whole idea. I mean, it's one thing to go to K——and EARN a slim body and good muscles. It's another thing, passively, to be cut and made prettier. I found it repugnant.
Mm-hm.
A girl friend of mine did it, and she sort of paved the way for me. And then I thought, "Well, it's silly not to."

This little vignette was in response to the question, "Have you ever had any operations?" Not, "Please tell me about yourself." Yet, in a few seconds she managed, while professing admiration, to be critical of the surgeon. While saying she admires people who, by their own efforts make themselves beautiful, she tells us she is not one of those people—rather, she does things for which she has revulsion. But what is one to do with such information? It should alert the listener to inquire further. Of which physician will she most likely next be critical? You guessed! She was soon critical of me. If her self-criticism and criticism of others is likely to be a cause of problems in the choice of treatment, or if recovery will be slow (allowing room for further criticism), then it may be wise to address the issue directly. If it does come under discussion, these little fragments must be supplemented by other evidence consciously sought in support of your conclusion. In other words, this interaction should raise suspicions but not lead in itself to conclusions.

This next utterance is in response to the opening question of the history of a hospitalized patient. It was the key to the case.

When were you last entirely well?
I'm never entirely well because of my back.
Mm-hm.

When something is crucial to a case, as was the meaning this patient attached to her back pain, it will come up again and again. Remember, the patient is rarely trying to mislead you. On the contrary, this woman wants her doctors to know how important her back pain is to her. Unfortunately even repetitive utterances are often ignored, when they do not fit the physician's formulation of the case. This does not mean that you have to change your understanding of the patient's problem just because the patient keeps stressing some other element. But when you see it differently than the patient, an obligation exists to address the patient's concern, if only to reassure. (If the patient cannot be reassured, then attention has not yet been directed at the underlying premise.)

Many believe that patients frequently conceal the true reason for their visit to a physician until they are about to leave. But even when this is so, their opening utterance may reveal a clue to the true reason for seeing a doctor. Sometimes the clue is simply that there is another problem, as in, "Well, I suppose that I thought it would be a good idea to get a physical." The word "suppose" suggests uncertainty, and implies the existence of other reasons. You are not obligated at the start to go digging for those other reasons (if they exist) but merely to remain alert to the possibility. Before the visit is over, however, you should be assured that the patient's entire agenda has been covered as completely as possible.

These next two fragments are important to the issue of truth telling. In neither instance was the patient asked specific questions about truth telling.

"I'm the patient, and I want to know."

"I don't wanna be lo— I don' wanna be lied to."

It is obvious that we cannot know with certainty that those two patients want to be told the truth, no matter what. But it should be equally obvious, once they have said it, that if deception is planned, then the doctor must collect compelling evidence to indicate that the patient "really" prefers to be protected from the truth.

Physicians must pay particular attention to personal meanings that come from a patient's lived past and knowledge of the world.

When this next patient tells us that she has had "nervous break-downs," we must pay attention, even though the information was not offered while her physician was inquiring about her past history.

I'm awfully hard on myself. I always have been. I gue—
What do you mean by that?
Well, I don't know, I kind of whip myself. I've had a couple of nervous—
What?
—breakdowns, like.

What led to this information was the simple statement "I'm aw-fully hard on myself." When one hears something of this sort, rather than letting it pass, as one might do in casual conversation, it is appropriate to go deeper. That wonderful all-purpose question—"What do you mean by that?"—will do the job. Because patients who have had episodes of mental illness always dread a recurrence, they may fear physical illnesses, procedures, therapies, or surgery for reasons quite different than those of other patients. Virtually everyone wonders, as they anticipate pain or disability, whether they will "be able to take it." Those who have suffered from emotional disorders in the past, however, have the added fear that the coming pain will destabilize them. For this reason they may require special reassurance that they are bearing up very well, if that is the case, or they may require extra support or psychotropic drugs if they are not tolerating their sickness. This patient, who had metastatic cancer, became depressed during her illness. The psychiatrist she consulted felt that she had very good reason to be depressed, in view of her fatal illness, and believed she should not be treated. The precipitating event in this episode of depression, however, as was the case in the past, were family difficulties, not the patient's cancer. The depression responded well to treatment.

The modern hospitalized patient has often had past experience with hospitals. These experiences contribute to the patient's store of knowledge and can influence the meanings attached to present medical encounters. All too often physicians dismiss the patient's knowledge as either trivial, unimportant, or even worse, unfounded. I overheard a patient, who had been repeatedly admitted for in-travenous chemotherapy, start to tell the house officer about previous difficulties with her veins. Like all such patients, she was very aware of the importance of preserving patent and accessible veins. She wanted him to wait for the IV team. The intern attempted to reassure her by telling her how good he was at venupuncture and then, predictably, failed. As this next example will illustrate, patients' previous experiences should not be dismissed. To disregard a person's experience is to disregard the person.

Optometrist.

Optometrist, yes. I was going on a trip to Florida vacation; I wanted to get sunglasses for driving. So I went in, and he looked at my eyes. And he said, "Your pupils don't react to light." And he said, "It wasn't that way the last time you were here two or three years ago." He says, "I advise you to see—your general doctor." He says, "Go to your general doctor," which I did go. He looked at my— and I had no symptoms of anything. No si— and I didn't even know it! No symptoms, and I went to the general doctor; he looked at it, and he says, "I don't know what's wrong, I'm going to send you to an eye specialist." He sent me to an eye specialist. He sent me to an eye specialist, right in A——where I live, and these two clowns got a hold of me and they were lookin', droppin', lookin'. And then they came back to me and they said to me, "It looks like you have Argyle Robinson pupils." Which in layman talk, he explained to me, meant, advanced stages of syphilis. And he says, "Well, we want to put you in the hospital and give you a spinal tap." He said, "That will tell." So my wife said, "Why you foolin' with these guys. Go to New York Hospital and get examined." Again, she got in touch with her doctor, who took care of her, and he recommended a neurologist. So anyway, he examined me, and he said, "I'm going to put you into New York Hospital and let them do a series of tests on you." He says, "I really can't tell what it is." So I went over there, and they gave me NUMEROUS tests, of X rays of my head and all kinds of— This was in June of 1972. They gave me numerous tests, you know, psychiatric tests with blocks and answering questions and riddles, and all that kind of stuff. And then they x-rayed me from head to toe, and that morning they said to me, we're going to do a spinal tap. They sent in a student, as far as I was concerned. There was a group of students that go around, and he was having difficulty, 'cause I had a spinal anesthesia once. And there was difficulty with the needle. Hurting me up a lot. And I was again leary of it, and then all of a sudden—'cause then he said you lay for four hours, he said, and then you can raise your head a little bit. All of a sudden they come in with a stretcher to take me for X rays. So I'm screamin', "I just had a spinal tap!" The guy says, "You don't have to get up, they'll x-ray you layin' down." I said, "This is insane!" So like a schmuck I got out, let them put me on a stretcher, they wheeled me down, I got down into the — layin' out there, there's paper on the stretcher—got inside, and this little tiny woman says to me, "I can't x-ray you, you got to stand up!" I said, "I just had a spinal tap, maybe an hour ago." She says, "Well, you have to stand up," she says. She said, "Hour's all right." I get up. Next day they said to me, "You're going to go to see our eye specialists." So I go up there, and I see him. And he examines me back and forth. And he gets my glasses out, and he says, "Your eyesight is not bad. In my opinion, you have Adie's syndrome, which is a nonmalignant-type thing." He says, "You could live to be a hundred; your pupils just don't react. Your pupils just don't react. This could be hereditary factor, or other things, but it's nothing to worry about. If I was you, I'd get out of here. Because you have been gettin' a lot of tests." So I come downstairs all happy, contented, you know; I'm going to go home. And three or four of the doctors come in and say to me, "Mr. C———, we got bad news for you. Very bad news." So my heart— this is ridiculous, I just come from the doctor who said— He said, "We could go along with him, but we found three times the normal protein in your spinal column." I said, "What does that mean?" He said, "Well, it might mean a lot of things," he says. "A lot of act— could be some activity going on." He says, "What we wanted you to do is we wanted you to stay and get a pneumoencephalography." He says, "It's an unpleasant," (which I found out) "but we're going to talk to your Dr. W———. So anyway I went in to see the head doctor of the floor, and I says to him, "Doctor, don't you think you should give me a little rest? You've been givin' me all these tests. I'm only a layman, but don't you think it's nuts to take this big test right away? For my own benefit? Or," I said, "why don't you do the spinal tap over?" I said, "Doctor, it was a very unpleasant experience, and maybe he did something wrong?" "No, no, no," he said, "This is conclusive. No, don't worry about that, it'll go right through." He says, "I'll tell you what we'll do. If you promise to come back, you can go home for ten days." He said, "We're afraid you won't come back." I said, "Look, I want to find out MORE than anybody here what's wrong with me, IF there's— I don't feel anything wrong." So I went home, gave my wife power of attorney over the business, you know, so many people crying and relatives, you know, like I was goin' to the death chair. I go back in, the same lousy tests over again, with the things and the questions— I said, "I already HAD all this!" But to me, in my layman's opinion, they were practicin' on me!

What should be noted in this example, is not only how distrustful the patient became (for good reason) but how compliant he was despite his distrust. Physicians have a special responsibility to heed what the patient says about the past because, despite their misgivings, most patients do what they are told. A long-term insulin-dependant diabetic patient began noting an increase in her insulin requirement coincident with local reactions to the insulin injections. She concluded that she had become allergic to the insulin. She was admitted to the hospital to bring her diabetes back into control. A full week was spent proving that what the patient said was correct. Why were her conclusions disregarded? Here, as in other situations, the patient is considered an unreliable observer. As ever, the Golden Rule is a useful guide. Physicians would be angered if someone dismissed our perceptions of events; our patients also become annoyed, or worse. When doctors must act in a manner that disregards the patient's perceptions or experience, an explanation of the reasons is warranted.

The patients' knowledge is employed to give meaning to their symptoms, and it is their interpretation of reality that causes them to act, not the reality itself. In these next two examples, we see this knowledge at work.

Actually, for myself, first I thought it was a pulled muscle. But then when it took so long to go away ... Then I was beginning to watch for, like a shingles kind of thing. Ye know, an' I thought. It just seemed to go ... because it was spreading ... and goin' across my back. An' I thought, well maybe its, like, following a nerve trunk. You know. I never saw any rash, or anything. Ye know, I mean its, its nothing that was serious. I just felt like you just shouldn' have a pain in your ... You should be able to lay on your left side without its being, having to move. And a' ... then I thought of neuralgia, and so on ... I kept ruling out all these things. Then I began to think, "Well what if its something really serious." An' I began to try and remember how you could diagnose Hodgkins and all these various malignancies. An' thought, "Is there something in my ribs an' in the bones 'n' so on, an' so forth ..." And thats what made me pursue it.

Note carefully, this patient did not follow up on her pain because it hurt, but because she began to think of Hodgkin's disease. In other words, the presentation of her illness was brought on by the combination of pain plus meaning. Let us conclude that if illness is a combination of a bodily symptom plus meaning, then effective treatment should also work on both the body and meaning.

Here is a similar example of the patient's knowledge endowing symptoms with meaning. In this instance my questions brought out the chain of logic that lead her to believe she had renal disease. Usually the system of meaning remains below the surface.

Ah, and then the other thing connected with it. I realize this in retrospect only. I've been doing this all the time. Strrettchingg. As if I had a pulled muscle or something. An' I go to exercise class once a week. An' I noticed that when bend to left, I felt weak on that side. But I reeely didn' pay any attention to it. Because I could get a puuuled something. An' Suddenly, it began to hurt HORRIBLY, an' I realized it was my Kidney! It wasn' my back at aaall. An' it hurt verrry badly.
What made you— what made you to realize that it was your kidney.
An Acute pain here.

How do you connect them?

I connect it with Sansert. Now I may be Absolutely wronng, but its the only, eh . . .

How long have you been on the Sansert?

About two years. A year an' ahalf, two years.

Now howdya connect these ankles and your kidneys?

Only that it has to do with water, somewhere. I mean its compleetly amateuur . . . commonsense analysis. Somepin', somepin' being retained somepla . . . I don't know why, I shouldn't even presuume to answer that, I don't know the answer. Only that Goodgold did tell me I suddenly remembered, that, a', he read me the textbook thing. The contraindications an' all that an' he said the only real danger is . . . for even have urinalysis is there is potential damage to the kidney.

Right. Now . . .

An' that suddenly came back at me, an' from then on . . .

Right. Up until, an' not counting these three events, ankles-knees-kidneys, and up until this week have you been feeling well?

Yes. An' let me tell you, the Sansert works. When I go off it. That's a funny thing because I had that for many years, but it was never diagnosed. I had electroencephalograms, I had what I call the Betty Davis test, the Dark Victory test . . .

If it is the meaning of symptoms, rather than the symptoms themselves, that cause people to visit physicians, why has the profession been able to escape dealing with those meanings until now? Many circumstances have brought about the change. Among these are the increased prevalence and importance of chronic diseases, increasing education and knowledge about medicine among the general public, a growing distrust of physicians and their technology, the rising cost of medical care, and the increased effectiveness of modern medicine with concomitant rising expectations among patients. Whatever the reasons, these meanings play a vital part in the presentation, course, and treatment of specific illnesses, whether or not a particular patient is aware of the contribution of his or her personal meanings. The patient in the previous example noted a series of events in her body. When this occurred, it called to mind something she had been told four years earlier. The present happenings, in the light of the earlier knowledge, led to an interpretation which, though incorrect, is certaintly understandable. A person's store of knowledge is part of a person, like their wardrobe. There is no way of getting rid of someone's store of knowledge without completely changing that person. It is obvious that one's store of knowledge changes through time, but so does one's taste in clothes, face, skin, diction, gait, habits, and virtually every aspect of a person; even the past changes, insofar as interpretations of past events change. Such change may occur in response to events of life—the same happenings that may lead to illness and disease.

But is the influence of meaning on illness "psychological" or "psychosomatic"? Most often when psychological factors in illness, or psychosomatic illnesses are discussed, what is implied is a direct effect of the mind and emotions on the body. Though I do not wish to deny such influences, the mind-body dichotomy as implied by such beliefs is, I think, old-fashioned. The mind-body "problem" is an unfortunate artifice of a particular historical context: that of the seventeenth century, during which time the partition of mind (or soul) and body served to guarantee to science a domain—the physical world— secure from the meddling Church. Hence the question of characterizing the influence of meaning on illness as psychological or psychosomatic is itself not sound or legitimate.

In the previous example something took place in the patient's back, and pain was perceived. This is the original symptom: pain in a certain place. As physicians we are usually more interested in what caused the pain than we are in what the pain causes. If we know the cause of the pain, we know the disease, and if we know the disease, perhaps we can treat it and prevent further difficulties. The desire to know the cause of things is universal. Although it is clear that illnesses do their damage because of what they do to the person, illnesses not only cause pains, swellings, dyspnea, and the remainder of the catalog of possible physical symptoms, they also cause patients to think about symptoms. Thinking about symptoms, attaching meanings to them, searching for explanations, are as much a part of the illness as are its physical expressions. These thoughts are not *caused by* the illness, they are *part* of the illness. Clearly an illness without cough is different from an illness with this symptom; similarly an illness in which the patient suspects or is afraid of, say, kidney involvement is different from an illness without such a fear.

We frequently forget that illnesses occur over time. Illnesses are processes, not merely events. In this process one symptom may lead to another. For example, if anorexia is the symptom, then weight loss and change in bowel habits may occur. And if anorexia lasts long enough, then malnutrition and its sequelae will follow. When a leg is injured, the gait changes; following that, the back may become symptomatic. And so it goes for virtually every disease. The fact that one aspect of an illness influences the next seems reasonable. So it is with meaning. Part of the illness is the meanings generated by the symptoms, and meanings compel people to action, which in turn may change their lives and alter the course of their illness or its treatment. Thus it should be clear that when meanings influence the presentation, course, and treatment of a sickness, they are not "psychological" factors in the old-fashioned sense. They are as much a part of the affliction as the physical symptoms. Seeing the two aspects, the physical and the nonphysical, as parallel and interactive is much more *useful* than maintaining an antiquated mind-body dichotomy in which the mind "does things" to the body. This view of the relationship is sometimes called psychophysical parallelism, but this too is a poor formulation. The physical and nonphysical are interactive, and if they were strictly parallel, they would not interact.

Some of the meanings that arise with illness are "unconscious"— not within the patient's awareness. These unconscious meanings may evoke emotions that have physical concomitants, which lead to further symptoms and further meanings, both conscious and unconscious, and so on, ad infinitum.

Because bodies are in many respects similar, we expect certain physical dysfunctions to express themselves similarly in different patients. Acute cartilage injuries of the knee cause characteristic symptoms in most patients. Equally, certain beliefs are so common that they should probably be presumed to be present even when unspoken.

I could drink water?

Sure!

Oh, I'd love to drink wata' . . . I've been dyin' to drink water.

Drink all you want.

OK, OK. Because I was afraid to drink it when I was home 'cause my ankles swoll up.

Uh huh

What was— what was the cause of my ankles swellin' up?

Your liver's pressing on the— on the big vein that comes from the legs an' jus' makes 'em. But water has nothin' to do with that.

I see.

You don't retain water because you drank it. An' if you don't drink water, you'll get yourself sick.

I see. I tried to stay away from all liquids so that my ankles wouldn' swell.

Nope.

In the same way fears of impotence and sterility are virtually universal.

_____

The problem with the prostate, the frustrating thing, is that there's not much you can do about it.

Well, yeh but, why does it make . . . what does it make you think is gonna happen?

That I'll eventually have to have it removed.

And what goes . . . What does that mean? I mean, you've heard of people having their prostate removed. How old are they?

Well, my uncle had it out about three years ago.

How old is he?

Ehhhhhhhh, late fifties.

How old are you?

I'm, ehhhhhhhh, thirty.

You mean you're worryin' about the operation you're gonna have in your late fifties?

(Laughter) . . . Well its a, eh— Well, no I eh— I mean, he didn't have a problem until his late fities . . . here I am thirty, no twenty-nine, and I'm having a problem.

Well, suppose you had your prostate taken out, what would that mean to you?

Ehhh, I don't know, eh; it might mean anything from sterilization to, you know, loss of having— loss . . .

Listen to me, you're not gonna' have any of those problems. When you think about it, what is it you think about?

Well, the prostrate . . . from what I know, supplies the fluid to get the sperm moving into the, ehh, into the cervix.

Well when people have their prostate removed, they really don't have their prostate removed. They just have a benign tumor on top of the prostate removed, they still have their prostate.

And it still furnishes the fluid for, ehhhh, having sex.

Yeh, yeh. You mean if you didn't have that fluid you wouldn't have sex?

No, it, it could be that you couldn't . . . wouldn't be able to have children in a normal way.

Uh-hm. Or that you'd be impotent.

Possibly, yeh.

So that— Is the fear sterility or impotence?

Mostly sterility.

Do you know anybody who got sterile that way?

No.

Where do you think it came from. Because I'm interested in how people think about illness.

Ehhhhh, ya know, ya just hear about it— ya know, ya just hear, I guess, basically. Over, the same way ... Well I learned about sex, basically, just by listening and, ehhh, listening in the streets, I guess. Nobody really had any education in sex, ya know. No one tol' me what happens if you have a problem. Here I am, the latest information I have is that— is that stupid book, that's a best seller now ... which anybody could really write. It doesn' give much facts. Mostly it leaves you in the dark. So I don' know, I'm just guessing. Just really guessing. I have no real basic facts to work with.

At age thirty (give or take five years), one of the more common afflictions to be brought to doctors is prostatitis-urethritis in males and urethritis-vaginitis in women. Associated with these disorders are fears of frigidity or impotence and sterility. Unless these fears are elicited or addressed, "recurrent episodes" may be as much the result of the secondary meanings as of whatever infectious agent may originally have initiated the process. In young men every penile twinge or transient discomfort on urination may be interpreted as evidence of continued or recurrent infection. Unless the doctor is cautious, he or she ends up using potent antimicrobials to treat anxieties, and in the process, confirming the patient's worst fears. "After all," patients reason, "if there were no infection, why would the doctor give me antibiotics?" In a woman anxious about the state of her vagina, and fearful of continued discomfort with intercourse, vaginal lubrication is often inadequate. Inadequate lubrication during coitus furthers the mucosal irritation, and a vicious cycle is initiated to which doctors may contribute by adding locally irritant medications as well as continuing to hunt for an infectious etiology. So it can be that a minor disorder of prostate, urethra, or vagina escalates into a major disruption in a young person's life because of the meanings attached to it by the patient.

Physicians sometimes attribute their frustrations in history taking to the idiosyncratic nature of the information obtained, as though subjective reports always differ from one person to another. Although there are individual differences, such differences are also found in anatomy. Remember the variation in the kind of myocardial disease that may follow occlusion of, for example, the circumflex artery. The advantage of history taking is that one can uncover the basis for the individual's response with further questions. It is more difficult to visualize the circumflex artery. The range of patients' beliefs about antibiotics exemplify such individual variation. They include "My body may get used to them," "They're too powerful for my body to take"; conversely, "I didn't start them because he only gave me

enough for four days, and I know you have to take them for ten."
Here is another, rather strange example.

Are you allergic to anything besides phenobarbitol?
Well, I think I'm allergic to it. Maybe I'm not. I don't know. I mean,
I was sick when I had this here cold. I was taking antibiotics, and they
opened my mind to things I didn't want to remember so I had to stop
taking them.

Whatever meanings have been attached to antibiotics by the
patient, it takes little thought to see how those meanings might
profoundly influence the course and outcome of a bacterial illness,
insofar as they influence whether or not the patient takes the drug!
Discovering that a patient has overridden the decision to take antibiotics, the physician may feel that his or her authority has been
questioned. When questioned, patients may explain that, although
doctors are expert in these matters, the patients know best about their
own bodies! Whatever the reason, meanings attached to the antimicrobials changed the patient's treatment.

Here is another almost universal fear, whose presence may determine whether or not the patient has adequate pain relief.

Let me ask you a question. Did you have pain?
Not much.
Not much! Not much!
Yeah, well I had a Demerol.
How come you don't wear the other—
I took a Demerol—
You did? And did it feel good?
No— but—
I mean, the pain—
Two hours—
Hm? Relieved your pain?
Two hours—that's not enough relief—
Well, then take another one in two hours.
I don't want to get to be a drug addict—

That physicians often provide inadequate relief of pain has been
repeatedly documented. The reasons are not clear, but it may be that
they share their patients' fears. In these instances it is not merely
patients' wrong beliefs about analgesics that determine the course of

their illnesses but physicians' beliefs as well. What doctors define as disease, what is considered worthy of treatment, how it is treated, when it is treated, what are considered acceptable side effects, as well as desirable outcomes of therapy, are part of the beliefs with which physicians operate. Although they are often based on considerable experimental and experiential evidence, it should be remembered that the meanings attached to objects or events by physicians do not necessarily represent objective "truth." These meanings and beliefs do, however, unquestionably influence patients' illnesses, aside from the effects of specific therapy.

Here is another example.

OK. How about your pain control?
Well I been asking for pain killers.
And are they working for you?
Yes, they are. But I don't like to ask for too many.
Why is that?
You get addicted.
Ma'am, that stuff is there for you. Addiction is not going to be your problem. You'll have plenty of trouble. But that's not one of them.
OK.

Meanings can be changed, however, as is demonstrated by this next interaction with the same patient.

There's three words you said to me, and I keep them in my mind all the time: "Anticipate the pain. Anticipate the pain."
Right.
So I did that. I felt like I was going to get a pain here, and I— I quic— asked for the painkillers right away, you know? And gee, it was terrific. I didn't even have any pain at all.
Isn't that easier than—
It was terrific.

Let us return to the woman with thrombophlebitis who developed pulmonary emboli. Her case has enabled us to learn how many different elements determine the actions of a sick person *aside* from the pathophysiology of the disease. In the earlier example the patient's avoidance of adequate medical care was documented. What finally made this woman take action? This next segment shows what happened:

Did you get new pain last weekend? In a new place?

No, the only thing that happened since last weekend was that this old vein where I had thrombophlebitis once—   ,

Had you told them about that pain in your leg when you saw them the first time?

Yes. Yes.

Did they know about the thrombophlebitis on the other side?

Well, not at the first time, but then, this is what, Friday? I'd say about Monday of this week or Tuesday that— but it flares up very frequently when I have a cold and I don't feel well—then it subsided. It happened to be red. And I just wrapped it up in towels—

When did it get red, please?

I'd say Tuesday of this week, and it was very red for one or two days. And now it's not so red, but there are one or two— it's not bad—one or two hard spots, possibly.

(Later)

I just want to tell you something—

I'm cutting you off, and I'll tell you why.

OK.

I mean, you're a sick lady. Whatever it is, we know that. You really know that.

Yes, I know.

You also know that people have been fussing around, now, for ten days with this.

I know that, yes.

Now, I have a sense that I do not want to fuss, I want to find out what's going on, I want to move.

Yes. Yes. Because now I feel the pain is— I tell you one thing I want to tell you about the pain.

Mm-hm.

Sometimes I get a pain, a spasm, apropos of nothing. And it just goes right through. And then last night, for example, I wake up, and there's a spasm. And it seems to be vibrating. And the only way I can contain it is by getting into a tub. It's constricting and then—

Right.

I had a feeling that it's going— moving, maybe moving in the direction of the heart.

Right.

And that I—

Weren't you fearful?

Well, I'm here because last night I really felt fearful.

And what was it specifically that made you feel fearful? What were you afraid of?

Well, I was fearful because I didn't have much breath, because the

sensation was coming— see, all along it's been on the left side, but it hasn't been in the center. And then, now it seems to be moving.
And what is the significance of its moving to the center?
In my mind?
Mm-hm.
Well I guess I'm thinking of blood clots.

Did she seek help because the pain was so severe? Apparently not, even though we are all aware of how distressing pleuritic pain can be. Rather, "I'm here because last night I felt fearful." She had good reason to be frightened, but even there her conclusion, correct as it may have been, was based on the wrong evidence. Although this patient's avoidance of adequate medical care is somewhat exaggerated, her behavior has a universal basis—she acts on her beliefs. Incidentally it is common among laypersons to attach significance to the fact that all the symptoms are on one side of the body or another, or that symptoms have moved across the midline. Thus when you hear a patient speaking in this way, you'll find that a few questions will bring out the belief and also reveal any effect the belief may have on the patient's behavior. Remember that when providing a history, patients may fail to mention symptoms that are at odds with their theory of their illness. In the same fashion physicians often dismiss symptoms that do not fit their working diagnosis.

This next patient's fear of paracentesis is easily understood.

Is that a surprise to you? That you got fluid in your belly?
Yeah, I thought maybe it wouldn't happen again. No, it just frightens me, that's all.
Well, what frightens you about it? What's frightening about it?
Well, I feel each time it fills up, you gotta empty it.
Well, is it the emptying that's frightening?
Yeah.
You mean the actual physical doing of it that's frightening.
And then testing the fluid to see if there are any cancer cells in it.
So the fear is that somebody'll do that and find cancer cells in it?
Yeah.
Listen, B——, that fluid is probably related to that cancer, it just doesn't have to have cells from it. And you don't care about that; you care about what it's doing to you.
Right.

Sometimes the personal meanings attached to a symptom come from the distant past. In this next example, which I have condensed, the original history was a nightmare: a psychiatric hospitalization for what turned out to be infectious neurological disease.

Any other times in the hospital?

When I was eighteen, I had a very big problem. I had loss of speech. I had a neurological problem, and I was hospitalized a number of times.

Tell me a little about that.

Well, I was sitting in my classroom, in my first year. I have really a lot of anxiety in recalling this. I was a freshman in college. And I had my foot crossed just the way I'm in your office now, and all of a sudden it fell asleep. And I thought something terrible was happening, but I didn't know what so I told the guy next to me, "Would you come with me because I'm really not feeling well." And as we were leaving to go to the school doctor, I started to loose my coordination for walking, and then I lost— I wasn't able to speak. And by the time I saw the doctor, I was just mumbling like an idiot. And he just said to my friend, "Have him go home." And I started running a high fever, and we got a doctor in Brooklyn to take a look at me. And before I knew it, I wound up in the psychiatric ward of Kings County Hospital, which was a very terrifying experience.

Oh I'll bet. I bet that was terrifying.

Yeah.

And then since then you've had no problems similar to that?

Right. No problems. But everytime my foot falls asleep, I—

You just feel scared.

I feel scared. Terrified.

When a patient comes to a physician for what seems to be a trivial symptom, about which he or she is extremely anxious, both the patient and the symptom will commonly be dismissed. In the light of this patient's history, however, his concern is understandable. When a patient behaves in a manner that is not easily explicable, it is more effective to establish the basis for the action, than to presume that the patient is "anxious," "hypochondriacal," or fits some other label that removes the doctor's obligation to solve the problem.

In these demonstrations of the influence of personal meaning on illness, I have stressed how accessible such meanings are to questioning. In so doing, I have implicitly minimized the importance of personal meanings that are beyond the patient's awareness—so-called unconscious meanings. As I noted before, I have done this to emphasize how much information *is* accessible to the patient's awareness and thus directly to the physician. Nevertheless, many personal meanings *not* directly accessible, and nowhere is this more true than in relationships with one's body and with one's immediate family.

One of the most curious and inexplicable phenomena in medicine is when a person acts (or fails to act) in such a manner as to injure his or her own body. For example, all asthmatics know that dyspnea is a *very* distressing symptom, and one would think that they would do *anything* to avoid it. Yet chronic asthmatics will occasionally alter their medication pattern and produce serious asthma. Why do they do this? One may say, "Well, I'm just self-destructive." The same phrase is used to explain many behaviors, from cigarette smoking to overt self-inflicted damage. I find the concept "self-destructive" used in this manner biologically counterintuitive, and useless, except as a means of intellectually dismissing an inexplicable behavior. But individuals do behave on occasion as though they were at war with themselves. In such instances these odd damaging behaviors (which are really quite common) may result from one "self" being in conflict with another "self" within the same person. On occasion a person may behave as though the body is not *their* body but instead belongs to one or another parent. No one says, "My body is my mother," but certain behaviors suggest this interpretation. Very little is known about the relationship between people and their bodies, and the observable facts are open to many diverse interpretations. In the context of inadequate theoretical guidance, your own careful observations become even more important. Let me follow this with an example that had a bizarre ending.

What I want to know, is, I also have been bothered this week by the MOST tender skin on my breast. It has been so— at times, so painful to touch! And I hope it isn't another one of my "mishegoss" [craziness]? Was—

Was what?

Bringing that on. (Laughs)

You mean—

You see, that was where my mother was radiated, and was burned, and was suffering a lot of pain.

Tell me.

And when it started, I thought, "Mother of God, I hope I'm not doing that to myself."

No.

This patient and her mother both had metastatic cancer from the breast at the same time! When her mother developed a symptom, so would the daughter. This is what she refers to in this example. One day she called to tell me that she was completely unable to move in bed. When I saw her at home, she related that she had gone to the toilet and while sitting there developed pain in the hip of such severity that she could hardly make it to bed. Examination revealed the limitation of motion but not its cause. After days of refusing, she finally consented to be X-rayed at her apartment. The X ray was negative, and I was able to coax her back to activity. Initially I had forgotten about the coincidence of her disease with her mother. I subsequently discovered that just before going to the toilet where the pain started, her mother had called from the hospital, having just been admitted for a pathologic fracture of the hip!

This patient's disease progressed, and ultimately, because of early neurological signs, she was admitted to the hospital and received radiation therapy to L4. She had an excellent result and was comfortable and psychologically well functioning and stable when she went home. One day her husband called me to say that he did not know how to tell his wife that her mother had died. I visited the apartment to tell her. She received the news calmly, saying, "Well, what did we expect? My mother's had a tough time, and she was an old lady . . ." The patient could not have been more reasonable. That was Monday. Thursday, at four o'clock in the morning, her husband telephoned to say that something terrible was happening. She seemed out of her head, and was repeatedly vomiting, drinking water and milk rapidly and then vomiting again. When I got there, she and the bed were a mess of vomit and diarrheal stool. She would not listen to anything I said, continually demanding more water and repeating over and over "I have to finish the pattern. I have to finish the pattern." I never gave thought to what she was saying. By six or seven in the morning, and after phenothiazines and increased steroids (she had had an adrenalectomy), all was calm, clean, and neat. At two in the afternoon her husband reported that she did not look well. At four, she was dead. Later that evening, her son said sadly, "Isn't that just like Mom. She finished the pattern—the same as Grandma!" My patient and her mother went down the road of the disease together.

Relationships within the family are enormously powerful determinants about which we know very little. Stories such as this one (and almost every physician has seen or heard of similar events), only seem mysterious because of our ignorance of mechanism.

The next patient had a mastectomy, and in her conversations postoperatively she always brought up the terrible illness of her sister.

Well, it's just that I watched this with my sister, and I hate the idea of chasing it all over my body because she ended up dying anyway.

Well, Jane, I don't know whether you're going to end up dying of this "anyway," anyway.

Yeah.

But I do know one thing. Between now and your death, whether it's from cancer or whether it's from an automobile accident, a lot of time's gonna pass. You are not faced with widespread cancer. I'm not in the business here of teaching you how to die. That's not the business I'm about now. I'm in the business of trying to make you a lot more alive. And, we're not in the business of chasing something. You know, taking pieces off you while we're chasing— no, that's not our business. Whatever it was with your sister, we're not your sister.

No, but—when you've watched it happen once— You know, there are many times when— the reason she didn't kill herself was because she was so close to my folks. But I thought, "How cruel for her." It would have been so much easier to do it in one lump. I know as a doctor you can't do that.

Well—

But I can.

All right.

I mean, what if I just ignored the whole situation?

If you get sick, then I'll take care of it. Take care of you, not "it." That's an alternative. Before we explore it—which I'm perfectly willing to explore with you—before we explore that, can we have it all laid out?

Because I watched my sister go for cobalt, and I watched her get nauseous from it. And I watched the sleepless nights, and I watched the agony that she went through—

What agony? Pain you mean? You're not going to have pain.

Ah, yes, she had a lot of pain and terrible nausea.

You're not going to have pain. I can solve that. There's no need to have pain. You— you do that, I've been there more often than you.

But all I could think of along the way was, "Jesus Christ, why don't they let her go in one piece."

Well, "THEY let her." That's another thing.

Well, it was like she was being kept alive with tubes and medicine and—

Nobody's keeping you alive with tubes and medicine. You're just getting over an operation. You do not have widespread metastatic disease. Will you please not leap forward like that.

(Later)

That was another thing that was so sad to me when my sister died. My father kept saying, "What did I do to deserve this? Haven't I been a good Jew?" And I said to him, "What has it got to do with anything, nobody's doing anything to you. You're not being punished." But it was his baby.

You remember that too, though.

What?

Nobody's doing anything to you.

No, nobody's doing anything to me ... But I shouldn't be forced to stick around and just ... just be kept, put together with spit for the rest of my life.

Please! Please! I'm looking at you right now, you don't look like you're put together with spit. Now ...

(Later)

I want you to tell me now about Elizabeth.

Betsy

So I know about Elizabeth, Betsy. So I know all about it. So I don't hafta, ya know, say to myself, "Well what's the hidden agenda," again. Every time ...

I've been working this in therapy for ninety million years, and it sti ...

That's your therapist and you. My consideration of— is when I hear a symptom, I have to have some sense of "Am I hearing Betsy's or Jane's symptom." Am I hearing the interpretation of Betsy's or Jane's. Now, that's a real phenomenon. You can spend the rest of your time talking to your therapist about it. Your therapist and I have DIFFERENT goals. I am not a psychiatrist. My goal is to get you the most functional time you have whether you live or you die. Function. That's what its about. That's what its about. All right. Will you tell me please ...

... I told you already.

No you haven't. How old was she when the diagnosis was made?

How old was she?

Yeh.

Forty-four. No! She died at forty-four; she was forty-one.

And how old were you?

Um, twelve from forty-one.

So how was the diagnosis made?

She, eh, they thought she had phlebitis in her left leg ... because the leg was swollen. And it was only after some time ... of even giving her a rubber stocking ... they found there was lump behind her knee. And they operated ... They found it was a sarcoma; they wanted to amputate her leg.

(Later)

So they didn't take off her leg. They did give her the cobalt.
They didn't take off her leg. Did they, did she know her diagnosis?
Um-hm.
And then what happened.
Then, um, next it was in her bile duct, and they took out ... I don't know ... from the size of her scar ...
But what has this got to do with cancer of the breast?!!
Nothing??
Now you—I'm going to tell you something.
I know, but cancer is cancer, isn't it?
And girls are girls. And East is West, and never the twain shall meet. What are you talking about?! You know that ... I'm going to tell you something. We missed some of the tape [recording] when you first mentioned this. But I guarantee you that I thought she had cancer of the breast! And then even later on when you said "sarcoma," I thought she had a sarcomatous breast tumor. But NEVER ONCE did you make, any distinction between what you had and what she had. It's a totally different disease!

This is a prototypical example. One sister, in her illness, confuses herself with the other sister. Even the least psychologically sophisticated physicians are aware of the complexity of relationships among siblings. Jealousy, envy, feelings of inadequacy in the face of the other sibling's achievements, the need to do things so that the other sibling will be proud of the patient, the desire to uphold the family's honor or achievements, are just a few of the items of emotional importance that may intrude on a patient's illness. When the patient believes her sister died of the same disease and desperately wishes to avoid following in the footsteps of her sister and the sister's horrible illness, the patient may go to extremes to avoid chemotherapy, radiation, or other definitive therapy, preferring instead "megavitamins" or some other unorthodox treatment. Such problems may be avoided by defining the patient's present situation in the most concrete manner and thus distinguishing her disease from that of the family member.

Here is another example where the illness of a sibling intrudes on the patient's concerns and cannot be dismissed casually.

Do people with thyroid problems have heart— high blood pressure problems?
Not your kind of thyroid disease, and that's not as long as it's controlled. You control your—you keep your thyroid medication going and control your thyroid at

a proper dose, which is easy to do, you should have no— there should be no difference between your life and somebody who has no thyroid disease.

I always had heard that people who have thyroid problems are more susceptible to diabetes, and it happens like—

That is correct. Fifteen percent of the people with thyroid disease are said to have or may be found to have diabetes. But it would be— well, we'll see what your blood sugar is tonight, but I don't think it is a worthwhile thing to go looking hard for. You have no tenden— your brother has diabetes?

Yeah.

All righty. Then we'll see what your blood sugar is. I don't think it's worth pursuing your diabetes—that is to say, glucose tolerance, and so on and so on, like that. Although I'm happy to do that, if you want. Or I'll find out it is something that has no meaning to you.

Mm-hm.

But yes, indeed, people who have thyroid disease more commonly have diabetes than people who have not but, on the other hand, so do people who have a family history. And that's you too.

Right. I was just wondering if in the future I should check it out, or?

You can do. You can have a physical every year.

Well—

Number one and number two, if you want to keep diabetes from being a meaningful thing? Exercise.

OK.

You're weight is fine, so the word is exercise. Running, swimming, biking, that kind of thing. Exercise.

When I found that my brother became a ... diabetic was I think when he went on shots and pills and like that (snap) like when he'd had a physical six months previously.

Mm-hm. How old was your brother when this started?

Eighteen.

Well, he was a young diabetic, and young diabetics have more difficulty with their diabetes frequently than do middle-aged diabetics. But if your diabetes is important, we'll pick it up in the blood sugar. We will NOT pick up diabetes in the so-called, "latent." And we don't find that by stressing your body's ability to handle sugar. And that is, we give you a load of sugar, and then we follow your blood after that. I do that, but I don't see the point in it.

OK.

I mean, if you need that to live a healthy life—

No, no. It's just something I wanted to make— keep an eye on.

Yes. You move. You—

'Cause I'm the only one left in the family who's—

I didn't know your brother wasn't alive. No? He died of diabetes? When was that?

March 16th.

All right.

Well, see, I'm afraid I'm falling into that kind of pattern, you know, like thyroid happened to him. I mean, I know it wasn't checked, and I know it wasn't— but I'm being historical about it.

———————————

Initially, a long, tedious conversation. But at the end, we find out that her brother died of diabetes, hence her concern. When I first listened to this tape, in preparation for a lecture, I realized that, although I had not seen her in eighteen months, I knew this patient's exact dose of Synthroid, yet I had forgotten about her brother. If she called today and asked whether she should change her medication or alter the dose, I would remember the details necessary for the decision. I also remember other details that are relevant to her care, but I did not remember that her brother died of diabetes at a young age. The reason that I remember one kind of fact and not the other is that I have spent the last twenty-five years learning to memorize clinically important facts. If a psychiatrist referred her to me he would remember to add that her brother died young and that they were close, although the psychiatrist might not remember which disease killed the brother. I am changing, however, and so must you. The death of the brother, the kind of cancer the sister had, what happened to the mother, are clinically relevant facts for physicians who care for sick patients. With a little practice these facts will remain as much a part of you as, say, the patient's dose of insulin.

Sometimes the effect of a family history can be relieved with a few words.

———————————

No, and I'm eating very little.

Why, because you have no appetite?

No. I'm scared, you know.

If you're not eating because you don't have appetite, that's one thing. If you're not eating because you're scared, come on. Your stomach doesn't know you're scared. Eat something.

Well, I'm eating a little. I don't want to gain any weight. My husband—

What are you afraid of?

—is so happy with my weight loss. He thinks I look so marvelous.

What are you afraid of?

Afraid of what?

You said, "I'm scared," so what are you scared of?

Well, you have alleviated one of my fears—

What was that?

A colostomy.

Yeah.

You have frightened me a little about my lung trouble.

Why is that frightening?

Well, my father had it—

You don't have a cancer of the lung.

That's good.

That's easy.

Yeah.

All right.

Now I'm not scared anymore.

Is that easy enough? You don't have cancer of the lung, and you're not going to have a colostomy.

All right. Fine.

Now we are going to return to the woman with thrombophlebitis and pulmonary embolus, who put off adequate care for so long. In this interaction we hear some more of the meanings attached to events in her illness—in this case going to the New York Hospital.

Now let me tell you something about the New York Hospital.

I know it well.

A good hospital. People are NICE to you— but you're going to be in a different part of the hospital.

I know.

Everybody there will explain everything to you. If you think something is going on, and you don't know what it is, and you don't know whether it's a yes or a no, just ask. You're not an—

I know the hospital very well. My mother died there.

All right, well, this is a different part of the hospital. This is the part of the hospital where people get better.

There, exemplified, is another factor from the patient's past that enters the case. The doctor can be as glib as he wishes about its being "the part of the hospital where people get better," but her mother died in the New York Hospital, and this fact will color her current experience.

Many of the patients whose conversations illustrated this chapter had serious organic disease: two with carcinoma of the breast, two with carcinoma of the bowel, one with thrombophlebitis and pulmonary emboli, and another with hypothyroidism. I chose these examples because I wanted to make it absolutely clear that the impact of personal meaning on illness is as true for serious as for trivial illness, and applied as much to patients with organic disease as to those with psychogenic problems. In one group of patients, however, personal meaning *is* the sickness: those whose symptoms are psychogenic.

The next case illustrates the point. This young man had already been to several physicians because of throat pain. I cannot be certain, but I believe that they were aware that he did not have organic disease. The antibiotics were given more as placebos, I suspect, than as definitive therapy. Because of continued distress, he continued to see physicians. I have edited and condensed the transcript to bring the relevant parts together.

Sharp and stabbing.
Yeah. Just a very fine point of pain.
And you say, when did it happen?
It just—
After a flu shot?
Yeah, about two days afterwards.
All right.
And, after that—
Did it last long?
Yeah!
How long did it last?
Well, the pain itself, the sharp, stabbing thing lasted, I don't know, thirty seconds.
And then?
Well, then I felt the pain there. But it wasn't, like, the first time I got it. And I saw a doctor right away, the minute that happened, and I asked him, "You know, what should I do about it?" And he just said, "Well, ah—"
Wait a minute. Back up a bit. The pain lasted thirty seconds, and that left you with some residual pain?
Yeah. So, I went away for the weekend, and after that, it really got bad. Like, my whole chin, and throat became stiff—
Say that again. Outside or inside?
Inside. The muscles became stiff, and I was having a lot of trouble swallowing with— to swallow. I had to push down, sort of, with the muscles in the back of my head and neck, and use other muscles to push it because it's like—

Is it the same in the morning and the evening?

Yeah.

It doesn't matter.

No. Well, what I thought just by looking at it was, it was very red and there was sort of red lines going all the way back that I could see.

Mm-hm.

The second time he also took a throat culture, and the results were negative. There were no organisms. He gave me an antibiotic. And I took that, and I felt a little stronger. But it didn't help me on my throat. It just made me feel a little stronger, and I got some bad reactions to it.

What kind of bad reactions?

Over here—it made me nauseous, for one thing, at one point, and I kept feeling as if there was a tightening in the middle of my chest.

I am now hearing something that happened before it all started. Something preceded this.

Preceded what?

All the history. Something started before the sticking.

No. It just—

What were you doing the day of the sticking?

I know exactly— I was sitting down on a chair watching television and all of a sudden!

The day before?

The day before? Nothing.

Something.

No. No, because I was surprised myself, 'cause it just came from nowhere. The only thing that happened was I had a flu shot.

Some other connection. There's no connection in your head between this and anything either?

No.

You sure?

Yeah. You seem pretty sure there's a connection.

I hear a connection that I'm not hearing. It may not be a valid connection, but I'm hearing a connection that was made somewhere, and I'm not hearing the antecedent data.

(Later)

Are your parents alive?

My mother's alive.

How old was your father when he died?

He was sixty-five.

What did he die from?
Cancer.
Of the?
Esophagus.
What else!
_____

My insistence that some event must have preceded the pain derived from his description of the discomfort. The usual diseases of the throat do not present in that manner. Infection does not start with brief episodes of sharp and stabbing pain. If a foreign body were the source of the pain, the patient should have developed pain after a meal, or after some opportunity for a foreign body to enter his throat. The actual description is more akin to the pain caused by constriction or spasm of the muscles of the throat. If the symptom has an emotional basis, then it should develop in close temporal relationship to some event that is the emotional stimulus. One must be as rigorous in the search for the emotional origin of symptoms as for any other etiology. Where psychological causes are suspected, the psychological stimulus will virtually always occur in close proximity to the onset of the symptom. A young man develops abdominal pain, for which no organic cause can be found. The physician suspects an emotional basis for the pain, after hearing that the patient's uncle died a month before its onset. If this is true, why did the pain not start at the time of the uncle's death? If the symptom is in fact related to the death, what intervening event caused the pain to start when it did? (And also, what is the physical basis for the pain? Even pain whose cause is psychological has some physical correlates—muscles in spasm, or some other source of pain.) Thus it is not sufficient to tell the patient that symptoms are due to "stress," without identifying the stress. Remember, what is stress to one person may be an exciting challenge to another. Nor is it enough to suggest that symptoms are secondary to depression, without identifying the source of the depression. If chest pain is psychogenic, why did the patient develop it at this time and not another? If the patient is having anxiety attacks, why now? You may argue that the psychological source of the symptoms may be buried deep in the unconscious mind of the patient, repressed and inaccessible to even the most clever questioning. Although that may be true, the relationship between the symptom and its source is usually obvious on a superficial level. If the doctor cannot find the relationship between the symptom and an emotional source, it is not fair to the patient to suggest some repressed or deep, but not demonstrable, psychological problem. Patients too readily take the blame for their illnesses as it is. This case makes the point. After hearing that his father recently died of carcinoma of the esophagus, it is more evident why this young man presented with throat pain. We know that he developed a symptom similar to the disease from which his father died. Further we know that such events are common. What we do not know is why *this* patient developed his symptom and what it means to *him*. The answers to such questions must await a psychotherapeutic setting. We do not know the mechanism by which such personal meanings bring about throat pain. The information provided by this interaction, however, is sufficient to provide a possible source for the patient's symptoms, provide a basis for giving him relief and discussing with him the origin of these symptoms (which effectively stopped his hunt for medical care), and provide the basis for a recommendation to a psychiatrist, which he followed.

In chapter 2 I showed how careful questioning could bring to light the pathophysiology of the illness—how the disease expresses itself at the whole person level. There I stressed the importance of not trying to make a diagnosis in a classical sense: fitting a disease name to the pattern of symptoms. It is better to bring to light the process of dysfunction occurring over time in order to understand what is making the person sick. This understanding will usually provide a disease diagnosis as a by-product. Of greater importance, however, is the fact that understanding the pathophysiology of *this* illness will yield a basis for therapeutic action that is more solid than is contributed by knowing the disease diagnosis alone. Diseases expresses themselves differently in different individuals. Coming to the diagnosis from knowing how the disease behaves in *this* patient allows therapy to be (where possible) more individually tuned. In the last chapter and in this, we have seen how personal meaning interacting with the organ dysfunction produces the illness, as the person experiences and acts on it. It should be clear that just as understanding the pathophysiology of *this* person's sickness aids in planning treatment, appreciating what the illness means to the patient, and how it has influenced the patient's behavior, should increase the patient's compliance and acceptance of the treatment regimen. But there is more. Knowledge of how the musculoskeletal system works is used in rehabilitation medicine to help patients return to function. In the same way knowledge of how personal meaning plays a vital part in the creation of the phenomenon we know as illness—turning organ dysfunction into a person's illness—should provide the basis for using personal meanings to make people better. In the next chapter I shall show how information can be used as a therapeutic tool.

References

Cassell, E. J. Reactions to physical illness and hospitalization. In *Psychiatry in General Medical Practice*. Ed. by G. Usdin and J. M. Lewis. New York: McGraw-Hill, 1979.

5

Information as a Therapeutic Tool

In the last decade, or so, there has been a change in the attitude of the medical profession toward informing patients about their diseases and treatment. Where previously painful facts about the seriousness and prognosis of an illness (particularly cancer) were concealed, now the full truth is more often spoken. This change in behavior parallels changes in the relation between doctor and patient. More often now the patient is considered a partner in his or her treatment; someone who must maintain autonomy, to the degree possible, in the face of illness. In the service of that autonomy, the patient is often believed to have an (absolute) right to the truth. The current demand, then, that doctors be truthful with their patients is primarily a moral imperative.

It is important to realize that in years past, when physicians were not truthful with their patients, neither they nor others considered the doctors to be "liars." It was widely believed that telling patients the terrible facts of their diseases was bad for them; the truth would harm them. I remember this dialogue from almost twenty years ago: the surgeon had found an inoperable carcinoma of the stomach, and when asked by the patient what the operation showed, the surgeon said, "Well, George, we did a lot of cuttin' and schnitten and pulled out some bad stuff, and you're gonna be okay." I cared for George until his death, some months later. The day before he died, and *only* the day before he died, he said, "Ya know, Doc, I sometimes wonder whether I'm getting better!" That was not an uncommon occurrence in those days. We do not know, however, whether patients were being any more truthful with their physicians than vice versa. I remember too well instances where patients demanded the truth and then, after being informed, became depressed to the point where caring for them was extremely difficult. McIntosh in *Communication and Awareness on a Cancer Ward* examines this issue as it arises in a British hospital.

Although he wrote it only a few years ago, the attitudes of patients and physicians toward informing patients are very much as they were in the United States a decade earlier. The essential fact is that physicians and patients usually share the same attitude about the truth—concealment is not simply something done to helpless patients by "paternalistic" doctors.

I would like to reiterate that twenty years ago in the United States "truth telling"—providing accurate information concerning terminal illness—often led to a depressed, worsened patient, with unmanageable symptoms and a destructive life situation. If information can do such harm, perhaps it can also do good. The moral imperative to tell the truth, however, seems an insufficient guide to what you should tell patients. It is difficult to argue against being honest and truthful, but merely characterizing a statement as "truthful" does not describe its content. What is the truth? It may be the truth to say, "You have cancer of the rectum," or "You seem to be failing rapidly," or, as I once heard said in response to the question "What is the matter with me," "You have multiple sclerosis, and you're pursuing a downhill course." Each of those statements, however true, somehow seems inadequate at best. The reason they are not useful, I believe, is because they contain very little information. If I tell another physician that my patient has carcinoma of the lung, and then ask advice about treatment, the other doctor will require much more information before he or she can respond: what type of tumor, where in the lung, the age of the patient, previous treatment, what other diseases are present, location of the patient (hospital or home, rural Georgia or New York City). When we say that a woman has carcinoma of the breast, three years after masectomy, the present tense is warranted because it conveys our understanding of the prognosis and what is required of her physicians today. The words, "cancer of the breast," have utility only to the extent that they convey information. It is obvious that the information carried in the simple declarative sentence, "She has cancer of the breast," resides not in the sentence, but in the listener (and the speaker). For example, for most of us the sentence, "She has Parrot's disease," although linguistically similar to the previous statement, contains remarkably little information. It does not even necessarily imply that the female is human. Thus it is possible, even likely, that the physician who said, "You have multiple sclerosis, and you're pursuing a downhill course," thought he was conveying information to the patient. It is unlikely that the patient attached precisely the same meaning to the sentence as did the physician, however, and thus it is unlikely that the doctor's understandings were transmitted to the patient. In common parlance, we would say that the patient may have misunderstood the doctor's words, even though they were truthful. Such misunderstanding is common, and difficult to avoid, since one cannot know exactly what interpretation a listener will attach to an utterance. One cannot know, because the listener's interpretation depends on his or her store of knowledge, physical condition, emotional state, the context of the utterance, the relative status of the speaker, the listener's needs, and much more.

How are we to solve the problem of understanding and misunderstanding? Perhaps the best way is not to look for evidence of the listener's "inner states," or understanding of an utterance, but for evidence that the utterance has accomplished its *purpose*. We must look, then, at the function of information. What does information do? To go back a step, we can safely say that one thing that all of us have in common is the need to act. In the medical arena, all diagnoses are followed by action, whether it be surgery or despair. Going to the hospital is an act; crying is an act; joy, sadness, suicide, chemotherapy, and virtually everything you can think of (including thinking) are acts.

The possibility of action almost inevitably implies choice. (Shall I laugh or cry?) And choice inevitably implies uncertainty. Where no uncertainty exists, there is no true choice. The opposite is also true: where there is no choice, there is no uncertainty. The problem of uncertainty in medical practice is a very important one because medicine is a probabilistic pursuit. Virtually nothing is certain, and yet, because it is a profession of action, physicians must act. Two fundamental functions of information, then, are to reduce uncertainty and provide a basis for action. The question is no longer whether to tell the truth but rather what the physician wishes the information to do.

To be useful, information provided to a patient must meet two basic tests. First, does it reduce the patient's uncertainty, now or in the future? Second, does it increase the patient's ability to act in his or her own best interests, now or in the future? When patient and doctor communicate, it is always in the context of their relationship; this relationship is the very heart of medical care. I believe that the relation between doctor and patient is the primary vehicle of medical care, and everything else in medical practice is secondary to it. Therefore it is reasonable to apply a third test to the information supplied by the physician: does it strengthen the doctor-patient relationship, now or in the future? Note that it is not necessary that information fulfill its functions in the present; what happens in the future is also a measure of its adequacy. Indeed, this may provide the only measure of the adequacy of the information provided by the doctor. In the simplest terms a patient may become furious on first hearing something, only to realize its importance later.

It follows that, if information wisely used can reduce uncertainty, improve the ability to act, and enhance the doctor-patient relationship, then information poorly employed can increase uncertainty, paralyze action, and destroy the relationship.

Here is an example.

Mrs. Black, Kevin's got a little murmur here.

Yeah. That's what the doctor in Chicago told me. But—they never told me what that was from, or what could happen from it.

Hmm. OK. Well. it comes from the heart, the murmur, and that kind of murmur— um, well, th— the noise the blood makes coming— coming through the heart. Well, when it's like this, when it's not exactly right.

Yuh. Well, that's pretty much what the doctor in Chicago said, but

Mm-hm.

I sort of wonder if it's serious—

Mm-hm.

—and I guess I just really didn't understand it. So if you were to explain about it to me I'd just feel better.

Okay. Well, it's really very simple. There's a little membrane that comes down and up in the upper chamber; there's a membrane that comes down. You know, ah, one from each direction. And sometimes they don't quite meet so— so that there's either a hole at the top or a hole at the bottom. Um, it hardly ever causes any trouble.

Oh! . . . S-sure—

The example does not meet the tests of reducing uncertainty, providing a basis for action, and strengthening the doctor-patient relationship. The doctor is attempting to explain to the mother what is the matter with Kevin's heart, but he is so unclear that he fails. He has not reduced her uncertainty nor provided a basis for her future actions in regard to Kevin's treatment. This example makes a further point: the information physicians provide to patients serves its function, in part, because it gives meaning to events. The word "meaning," as used here, has two primary senses, or connotations. The first sense, is that of *significance* or *implication*. The meaning of Kevin's murmur is that it signifies atrial septal defect. The second sense of "meaning" is *importance*. Information or events have meaning to the extent that they are important to someone. It is obvious that, although the same events may always signify the same thing, their importance depends on many other circumstances. The concept of importance implies that people are involved. The terms "value," "concern," and "care about" also convey the sense of meaning as importance. Even if Kevin's mother was able to extract the significance of Kevin's murmur from what the doctor said, would she have known whether to be concerned about it? Obviously not (despite the doctor's statement that "It hardly ever causes trouble," what does "hardly ever" imply?). Thus a medical explanation must contain more than simply the significance of events, it must also begin to answer questions about the importance of the events to the patient. I say "begin to answer," because importance is an individual matter. When we provide information, we must make sure that we are addressing the questions of that the particular patient, which may differ from the questions of the physician, or those of other patients. Another way of putting it is that a medical statement contains three parts: the facts, their significance, and their importance to the patient. Kevin has a murmur. Kevin's murmur signifies the failure of the wall between two parts of the heart to close, which is known as atrial septal defect. Such defects do no harm and do not require treatment.

No listener would think the preceding example represents good communication between doctor and patient. The next example, taken from a British book on doctor-patient communication (Byrne and Long 1976, 15–18), was cited as an example of the proper way for physicians to communicate with patients.

Come in! Good morning! Sorry to keep you waiting, Mrs. M——. These are two medical students.

I presumed so.

Please sit down.

After seven children, I've seen a lot of them.

A lot of medical students?

Yes.

So what can I do for you?

To be quite honest, I don't know. I just feel awful. I keep going hot and cold. My eyes come up. I'm nearly 45 and suppose I must be on the change now?

What shall I tell you now?

(Laughter) No . . . Women have done it before.

Is that what you are telling yourself?

That's what I'm presuming ... I don't know.

Mmm, mmm ... What does the change mean to you?

I don't know. But my periods are some months heavy ... some months I have two.

Mmmm, mmmm.

Last month I had two, and the one now was due last Friday.

Mmmm, mmmm. What does the change mean as an idea to you? I mean ...

Well, I don't know. I have an idea that it is usual for women. They have these hot flashes, and they don't feel for this, and they don't feel for that ... But my mother used to have blackouts.

When she was on the change, you mean?

Yes.

Mmmm. And are you worried you would be ...

Well, I don't know ... but I'm frightened.

Mmmm, mmmm. Frightened of what.

Of getting blackouts, or anything like that.

You are frightened of more than blackouts?

(Nervous laugh) Well, it's not that I choose to. We used to find her miles away on buses. (Started to cry.)

Go on—

(Crying openly now) Well, she never knew how she got there.

Mmmm, mmmm. And that was supposed to be due to the change?

Well, that's what at she told me. I don't know.

And when did she tell you that?

(Continues crying for a few seconds) Well it's years ago. She's dead now. I think she was going out of her mind. Well, it's years ago ... 23 years ago.

Mmmm, mmmm.

Well, she used to go out on the bus. Then they'd find her miles away. And she never knew how. I never knew, but my father used to say she was on the change.

So for the last 20 years, you've thought of the change as a time when you ...?

Well, I have really.

So, when you get to 45 and your periods get irregular ...

Well, you think. Well, you do get worried ... I suppose. I don't know.

But you know, because you've seen in the past 20 years other women get past the age of change without going the way your mother went? So does that mean you feel more like her than ...?

Well, you know, are these things inherited—that's what I want to know? That's what's worried me.

. . . .

It's just that I think about this um um change. God help I don't go like her.

Do you have any sisters?

Mmmm, mmmm.

Are they older than you or younger?

They are younger.

Have they got the same kind of fears about the change as you do?

I don't know. I've never asked them. One's 28, and the other's 36.

So you are a good bit older than them?

Yes. I can remember, and they can't.

There perhaps was a time when you were almost as much their mother as your mother was?

Well, it could be. Well, I almost brought the others up, you know.

Are you a motherly person?

Well, I've got seven of my own. (She has stopped crying and is now beginning to laugh again.) So I am . . . yes.

Have you talked to anybody or your husband about this yet?

No, I haven't.

So you've bottled it all up. And now it just won't stay bottled anymore.

(Crying again.) No, but it's silly . . . Nobody ever asked me before, before you asked me. I know it's there when I think about it.

. . . .

Do you feel any different now, from when you came in?

I don't know. It's a silly thing. You have seven children and you do this, and you do that. You've got to get it behind yourself. You pretend it's not there, but it's there all the time. And you keep thinking to yourself, well, it might be all right. I've been happily married, why worry!

So today, you've done something very different from what you normally do?

I have, yes. I never expected it.

Right. I think that's very good. I'd like you to come back and see me next week.

(Still crying.) Is it the change?

If your periods have been going irregular then it may be. Sure.

Well, I've always been regular, let's face it . . . As far as these things go. Now it goes that I have none this month and then next month I have two.

Okay. We can go into those sort of details next week. But you've done something very big this week. I don't want to start confusing you now with all sorts of little details.

This technique of reflecting back to patients what they say, allowing them in a sense to hear their own questions, is frequently employed by psychotherapists. The patients are given an opportunity to ventilate their feelings. Although this exchange is cited as an example of good communication between doctor and patient, I find it unsuited to the situation. It is unsatisfactory, in part, because this psychotherapeutic technique is being used for purposes other than that for which it was designed. I believe the problem presented by this patient at this stage is quite simple. She has certain symptoms and wonders whether they imply that she is starting her menopause. If she is menopausal, she wonders whether she will become as nutty as her mother was at menopause. These seem to me to be very reasonable concerns, which were not addressed by the physician. The patient is well aware that all women ultimately become menopausal, so this is not her question. The physician got the patient to verbalize her fears, which is to say he caused her to tell him the importance of menopause to her. But then he did not answer her question. You may say that he cannot answer the question, that he does not know how she will behave in the future and whether or not she will follow her mother's pattern. But there are very solid answers that can be provided at this point in time. First, menopausal, or not, the patient is not crazy now. Second, it does not follow from anything known either about the menopause, mental disorders, families, or inheritance that the patient will become mentally ill in the future. Third, though the doctor cannot promise that she will not follow in her mother's footsteps, he can promise that if she begins to have trouble, she can immediately come to start treatment and get effective help. In other words, the doctor can reorient the patient to the present and offer solid reassurance about her current situation. He can also reorganize her view of possible future events and then offer reassurance about them. Finally, the doctor can begin to determine whether the patient is actually entering menopause and so put the issue to rest. After the doctor has done all these relatively simple things (in much less time than was occupied in the example), the patient still may be obsessed with the fear that she will follow her mother's path. In this case psychotherapy may be necessary.

Physicians may view the function of the physician, in this setting, from different perspectives. I believe that each time doctors see a patient, we must ask ourselves what is the problem the patient is presenting. The formulation may include the diagnosis (in classical terms), but it is certainly not limited to the diagnosis. Every problem has possible solutions. If the problem presented by the (possibly) menopausal lady is seen as emotional, and possibly requiring psychotherapy, then any solution involving therapy must take into account the patient's willingness to participate, the requirement of treatment extending over time, and the possibility that other emotional areas requiring (or believed to require) therapy will be uncovered. If these issues are not considered, then why start a psychotherapeutic treatment? If, as I believe, the problem is that the patient wants "reassurance," then "reassurance" should be provided. I put quotes around the word reassurance to acknowledge its complexity and to make it clear that to reassure someone is not merely to dismiss them or their concerns.

"Don't worry" may be the most commonly employed and least effective expression of reassurance in the language. It is said that Dr. Cecil, the renowned professor of medicine at The College of Physicians and Surgeons of Columbia University, was walking down the hall of Presbyterian Hospital when he heard a student saying "Don't worry," to a patient. He called the student over and said, "Smith, I'm glad I ran into you. Your name came up at the meeting of the Promotions Committee this morning, and we had quite a lengthy discussion. I'm not at liberty to tell you exactly what transpired, but ... uh, don't worry." The expression is *not* useful because the only information it carries is that the speaker wants to reassure, and that therefore reassurance must be called for.

An utterance, or act, provides comfort or offers hope when it addresses itself directly to the concerns of the listener. To comfort, then, the speaker must have knowledge of these concerns. Some fears—of pain, for example—are virtually universal. However, simply saying, "Don't worry. This won't hurt," may not be effective, because it has the sound of cliché. "This won't hurt, because the instrument never touches your skin." "This will hurt briefly, like a small needle stick, and that will be the end of it." These phrases may be more reassuring, because they have sufficient detail to satisfy the patient's fears. Reassurance must have credibility. Everyone wants to believe that things are going to be all right, simultaneous with the fear that they never will be. We must guard our credibility if we wish to be believed in the future. If you say it will not hurt, it better not hurt! In the example of the woman concerned about her menopause, the physician seems to have obtained an accurate picture of her fears but still does not address them. Step one, of reassurance, is finding the nature of the problem; step two is addressing the problem accurately and credibly.

Throughout this chapter you will find examples quoted where you may disagree with what the physician says. You may believe that the doctor has the facts wrong and is giving the patient an incorrect explanation. Or you may feel that he or she is communicating poorly—talking to much, not letting the patient speak, using language that is too informal or filled with slang. These are two different issues, and both are important.

Strange as it may seem, for information to serve the functions that I am describing in this chapter, it is not necessary that it be correct. In fact most medical explanations over the years have not stood the test of time—new information often proves old understandings wrong. As Dr. John Coulihan has shown, the explanations of chiropracters to their patients serve many of the functions that are demonstrated in this chapter but employ an explanatory system with which most physicians are not in agreement. The basic point is that the information must be credible to the patient. Believability requires not only that words be plausible but that a speaker be worthy of trust. Both criteria are met best by someone (the physician) considered an authority by the patient, who is speaking what he or she believes to be the truth in a language that can be understood by the patient.

The issue of style is another matter. Every physician must develop his or her own style of communicating. Sometimes, reading these examples, I get irritated at the doctor. I disagree with this or that feature of the explanation, or with the vernacular quality of the speech. On occasion I am dissatisfied even though I am the physician in the example. Since these examples were recorded, I have become much more efficient, and my language is now more precise. In editing this chapter for publication I debated whether I should revise the examples—sanitize the language and ensure the (current) correctness of the explanations. I decided against doing that because I want the reader to realize that doctors are persons also. Physicians must develop skills to meet the needs of patients, but in the process of becoming more effective and efficient, they must remain true to themselves. If they do not, if they try to act a role rather than *be* a physician, then they will not, I believe, ever be the best instrument of care that they can be.

In the previous chapter we met the young man with persistent pain in his throat, who had been to several physicians without obtaining relief. His father, you may remember, had died of carcinoma of the esophagus eight months previously. As you go through this dialogue, you will see that reassurance is offered in several steps. First, I tell him that I think his symptoms are emotional in origin. Next, I suggest why they are emotional, obtaining his agreement as I proceed. I then explain how the emotional material might produce the physical symptoms. Then, I suggest a treatment I believe will relieve his discomfort. Following that, I review the evidence for his *not* having cancer. And finally, I make it clear that I will be there to help him if he does not get better. I also qualify my competence by specifically stating that I am not a psychiatrist (not "playing" psychiatrist). You will notice in what detail the pathophysiology of his symptoms is discussed, as well as how his emotional distress became translated into physical symptoms. It has taken me a number of years to learn the steps, because textbooks do not contain this information. Aside from my suggestion that localized muscle tightness is often responsible for physical discomforts of emotional origin, and that a persistently tight muscle ultimately becomes tender, you will have to learn the rest from your own experience. If you accept that physical symptoms *always* have a physical basis no matter what the underlying cause, you will quickly learn this step.

My guess is that somebody suggested to you that this is emotional in origin and that it's what's called globus hystericus.

Ah—

And that— that's not a sufficient answer to you.

Some— I mean, ah, nobody has really suggested it, but the thought did occur to me.

All right. I think that it is emotional in origin, but as I said, that is not a sufficient answer.

Mm-hm.

In other words, um, first of all, I don't think that it is— I'd like to set a chain up— a causal chain up for you.

Yeah?

You are well within the grief period for your father's death. Second, you have much more responsibility than you expected to have, much faster. And that is your mother.

Mm-hm.

And your brothers and sisters, while they may share with you the responsibility, have their own families.

Right—

OK, in the light— now we have that setting,

Mm-hm.

—that your father died of carcinoma of the esophagus, of the throat, in layman's terms, that's here, and you get acute symptoms. And the symptom acquires a meaning it would not have for her (a student is present) or for me. But meanings don't make it difficult to swallow.

I mean, you know—obviously. I mean, that thought did occur to me. I thought— my first reaction was, you know, "This could be cancer." And you know, I thought— that is one of the first thoughts that came to my mind.

Well, meanings don't make it difficult to swallow. In other words—

Yeah.

—no matter what you think—

Right.

—that doesn't make it hard. Unless somehow what you think is translated into the body. And the way things are translated in the body is with muscles. You've got the tightest— tightest set of throat and neck muscles and the tightest set of chest muscles of any young person I have ever saw.

Yeah.

But they are not tight because they're damaged. They're tight because they're tense.

Well, I'm the kind of person where, ah, you know, I hold in my, ah, my emotions. Like, I guess you can see I'm not the most talkative person in the world, and, ah, you know, I tend to hold things in.

I know that. You've got your chest held in, and you've got your throat held in like that. Tight muscles hurt. They hurt, and they make tightness. By the time it goes on a little while, they also make the mucous membranes in the throat sore because they're tight. The point is that this is very real. These symptoms are not imaginary symptoms. They're very real symptoms.

Right.

And they are self-propogating.

What do you mean by "self-propagating"?

Well, in the sense that their continued presence tends to continue their presence. Tight neck muscles make for sore muscles which makes for tight muscles which also makes a worried person above 'em.

Yep.

What would happen to you if you got—cancer of the esophagus.

Mm.

Um, what would happen to you if your mother said to you— if you got sick now?

What would happen to what—?

I mean, the whole—

There'd be a lot of grief.

Wouldn't it?

Yeah.

Now. If I could get your muscles loose for just two or three or four days in a row, you'd feel better. Your throat would get better. If I could get you to sense that the body is a complicated piece of machinery. There are a lot of ways around the body's relationship. Ways to loosen your muscles up without necessarily under-

standing how it all came about in the first place. And that is Valium. And we'll just give you three or four days of a relaxed body, until

Yeah.

—you get yourself back in your own hands. Listen, you just spent a hundred and some odd dollars on some very real tests.

Mm-hm.

And every one of those real tests is normal. You had a barium swallow: normal. If there's any question about it, get me those pictures. You do not have infection in your throat. You do have some postnasal drip—join the crowd. You do have a very tight chest. But it's muscles. You're breathing abdominally, so you're going to be short-winded. That's not your normal way of breathing with your chest—it's just held like this. And it may, for all I know, be you're holding in grief or whatever. But it's a psychological interpretation which is beyond my competence.

Mm-hm.

But what is within my competence is what the body does.

Mm.

I'll give you some Valium—not even very much. Take it three times a day. Go get some steam or a massage if you can and start working on that body. Work on that body! It'll do things for you.

Mm.

It really will. You call in a week. If you're not better, come back in, and I'll keep at you. I intend for you to be better.

Here we see how many functions are served by the doctor's information. The patient is told what is the matter, why, what is to be done about it, and is assured of the physician's interest and accessibility. Conspicuously absent from the doctor's discussion is a detailed psychological interpretation of the patient's illness. There are two reasons for this absence. The first is that the doctor cannot know the psychodynamics with certainty. Whatever interpretation is offered (some, who have listened to this taped conversation, have suggested unresolved Oedipal conflicts, while others have proposed pathological identification with the father), it may be wrong. Since the facts cannot be demonstrated, the physician may be in the uncomfortable position of saying "yes" and having the patient say "no," with definitive resolution impossible. The second reason for not offering psychodynamic interpretations is that no action will be based on them. The information presented should be used to do something in the patient's behalf. On the other hand, one could, if one wished, offer several alternative psychological interpretations, indicating that people sometimes have experiences and act in certain ways for reasons of which they are not aware. In this instance unconscious determinants in general are discussed, without the physician having to defend a specific interpretation. The patient's experience is legitimated: something is indeed "wrong"; the symptoms are real and are a bona fide reason for concern. Legitimating the behavior of people with respect to their health is one of the functions of doctors. "It's all right to be nervous like that." "It's all right to feel your penis, breast, or what-have-you." One of my patients was distressed at her flatulence. I pointed out that almost everybody passes considerable gas as they get older. "Why doesn't everybody know that?" she said, much relieved.

Information has specific features, each of which can be varied to fit particular situations. The first is the amount of information. Will the explanation be limited to the specific medical event—for example, an episode of cardiac arrhythmia—or will it cover heart disease in general?

A separate but related feature of information is the specificity of detail. Some patients require an explanation all the way down to the mitochondrial level before they feel that they understand their situation; others do not mind a rather vague picture of the problem.

Information involves timing. Is it offered preoperatively or postoperatively, in the recovery room or after the hospitalization? Is it provided at the beginning of the visit or at the end? Why, for example, did the physician not tell the man with the sore throat that his pain was psychogenic right at the beginning of the visit, when these suspicions were first aroused? When enough of the patient's history had been obtained, so that the physician was quite certain he could have said: "Why don't we save ourselves a lot of time and trouble. I think your problem is basically emotional: you've got some hang-up about your father's death, and this has translated itself into a sore throat." The reason for not doing this is simple. The patient would not be convinced. This doctor had not examined him or done laboratory tests. There had been no further discussion that might lend weight to the doctor's words. The same statements, at the end of the visit, would be more effective. Information has a job to do, and it should be employed in such a way as to maximize its effectiveness.

Information has meaning. As you are aware, from volume 1 and earlier chapters in this book meaning comes, not only from words and syntax, but from context, the expectations of participants, their roles, and the history of the encounter and the relationship. Let me emphasize that what counts is what statements mean to the patient, *not* to the doctor.

Finally, information has truth content: within confidence limits, it is true or false. It is difficult for practicing physicians to be as rigid as some others about the need for for truth telling in clinical situations. There are occasions when a lie seems the best thing, although such situations are infrequent. In general, the truth is best for the welfare of the patient, and everyone else concerned. There are certain practical problems with lies. For one thing, you must remember what you said. For another, you have to hope everyone else involved in the patient's care employs the same lie. Finally, when patients discover they have been lied to, their trust is undermined. The truth is better and easier.

The kind of information, its amount, degree of detail, timing, and truth value vary with the clinical situation. In view of all this the use of information may seem a complicated matter. It is no more complex, however, than the use of a potent drug, for which decisions must also be made about how much, when, and in what form. Unlike a potent drug, however, which you may chose to employ or not, information is generally always used by the physician. The question is not whether to use information but how well you are going to manage its use.

Uncertainty

As long as we are alive, we must act. In order to take action, however, decisions must made, and decisions always imply uncertainty. The more urgent the need to act, however, and the more threatening the situation in which action is required, then the more intolerable uncertainty becomes. Thus the greater the need for action, the more accurate must be the the information in order to reduce uncertainty. Unfortunately urgent situations are precisely when uncertainty poses its greatest threat to the doctor. At such times, when the patient most needs to know what is going on, doctors often become extremely vague—perhaps to protect ourselves from our own uncertainty.

In this next example the physician explains not only what has occurred but what is going to happen. The patient is being prepared for future uncertainties, with their attendant fears.

You had thrombophlebitis when you left Rochester, and you've thrown a clot to your lung. And you did that, ah, um, if it's only one, you did that when you got this awful pain; probably threw several small ones and then the big one. Now, let me tell you about it, and what it all means to you, and so forth. First of all, the clot that you have—that'll go away and leave you without any total disability or anything of the kind. Our worry is— our concern is not in, ah, in immediate terms, is not what you have now. As they used to say, "If somebody shoots at you, if you heard it, it missed you." If you're sitting there, it missed you. It's not that one that concerns me. It's the next one.

Mm-hm.

And so our whole problem is to prevent the getting another embolis. I mean, the overwhelming probability, like in the ninetieth percent probability is that you had thrombophlebitis and threw a pulmonary embolus, and that's what's in your chest now. And you've got some fluid from the pulmonary embolus, and your pleura is irritated from that. And that's why it hurts so much, and that's why you're short of breath. And our concern is that you do not have another embolus. And, ah, we do that by admitting to you— admitting you into the hospital promptly and anticoagulating you. Just like Mr. Nixon, we don't trust people to anticoagulate themselves at home. They don't do a good job. Now—

If I had moved, I mean, I'd— the way it moves—

I press this— I press this button to make a telephone call to get you a room at New York Hospital promptly. Now, I only want one thing—

I can't go home to get things?

Let me see if I can get a bed, will you? That's the first thing. Then we'll—

How long will I be there?

Huh?

How long do you expect I'll be there? Just roughly.

I'm going to tell you. I'm going to tell you exactly what's about to happen to you. I'm going to tell you everything we're going to do. I'm going to tell you everything about this. And I'm going to explain to you ahead of time only because some things will start before I get to see you again later on this evening. And I want you to know exactly what's happening, exactly why everybody is doing everything, and, uh, I'll discuss that in a minute.

Mm.

Here's what we have to do. As I said, the urgency comes not because of the pulmonary embolus you have now. That is not a threat to your life. It's a threat to your well-being and sleep, and so forth. Our immediate concern is that you be anticoagulated. Now— but we do some things first. You'll get another chest X ray, you'll have some blood work done, another person will examine you—the intern and the resident will examine you. And those things will all be done within the next couple of hours. And then you'll start on a drug called Heparin, which is given intravenously.

Mm.

A good drug. And it's a very effective one. And when I get there this evening, I'm going to give you a nerve block on th— in your back and see if I can't eliminate that pain or at least decrease it so that you can breathe better. Ah, I will be there at—

I can BEAR it, except a moment, you see—

I understand you can bear it, we already— you have already proved to us what a good bearer you are. Now just be a patient for a while, would you please. OK. Now, ah, after you are on Heparin for a couple of days. As a matter of fact, the minute your Heparin is really in and established, which is hours, my concern for your safety diminishes remarkably. And then I'm not worried about you anymore, and then we-we go ahead to make sure there is no other disease that might have gotten this all started. I do not believe there is. You look to me, from the rest of what I've seen, like a healthy lady. In any case within a couple of days, we start to transfer you from the intravenous medication to Coumadin, which is given by mouth and takes just a few days. Usually between ten days and two weeks. Sometimes less, and you're back out again.

Now, when I'm out, do I have to be caref—

You will— No.

Do I have to be careful for a while or any special thing?

You're going to find that you've been sick.

Yes.

As a matter of fact, you're already feeling so much better—

Yes.

—that it's no longer in your hands, it's now in somebody else's hands, and that's a very comforting feeling when you've been sick. And, uh, you will find that when you let go and get some sleep, and so forth, you'll feel better. When you get out of the hospital, you'll tire relatively easily, but you'll come back to yourself. You're a healthy lady.

OK.

The, ah, Admitting Office is reached through 68th Street and York Avenue. There's an entrance to the hospital right there. You just tell them your wife is going to the Admitting Office, hmm, and, ah,—

From here on it's luck!

Oh no!

No?

No. Up to now, it's been luck!

One might argue that the physician could not be sure that what the patient has been told will occur. Other physicians will be involved, as well as other hospital personnel. Indeed, the physician cannot be certain, in advance, of the truth of his statements. He is more certain than the patient, however, and she understands that the doctor is suggesting what will happen rather than making a pronouncement.

Here is another example of a doctor's explanation to a patient. This woman went to the Employee Health Service at Bellevue Hospital because of visual symptoms. The possibility of pituitary tumor was raised, and X rays were ordered. As in most big institutions, things take time at Bellevue. Therefore for several weeks the woman wondered whether she had a pituitary tumor, and what course of action would be taken. Such concerns would make most people nervous. The woman then took her X rays to another physician; this is their conversation. Not only are the patient's uncertainties addressed, but also the actions she must take are discussed. The effect is to strengthen her relati[,] ıship with the doctor. The information meets all three tests: reducing uncertainty, providing a basis for action, and strengthening the doctor-patient relationship.

Ok, Ma'am. I've got some— we've got some conversation to have.

Yes.

Now, first I want you to start off by listening— listen to me, carefully. With your ears, not your fears. I'm very straight with people, as you know. What I tell you is everything, uh— right. Now, on your—skull X rays there is a place where the pituitary gland sits. I'll show it to you. That place right there—

Mm-hm.

—is larger than it's supposed to be.

I see.

The pituitary gland sits in there. Now the pituitary gland has a lot— does a lot of work in the body. It's very teeny and makes a lot of little hormones that tell the other glands of your body what to do.

Mm-hm.

When it's large, like that, we think it has a tumor. The kind of tumor that it has is not a cancer tumor. That is not what makes that kind of tumor.

Mm-hm.

But it probably, not surely, but probably is a tumor.

Mm-hm.

That would give you the headaches that you complain of. 'Cause tumors cause headaches, they grow very slowly. They're not cancer. They grow very, very slowly.

Mm-hm.

They make their effect by just getting a little larger making pressure. But you can get the same kind of picture with no tumor. In other words, all that does is suggest. It doesn't tell you that it's there, it suggests it.

Mm-hm.

The reason that—makes—trouble with vision is that the nerves that go to the eye go right past it.

I see.

So if it's in between them, and it gets larger, it pushes on those nerves and that's why you get sometimes eye symptoms from that.

Mm-hm.

But you can get the eye symptoms you had, and there's no tumor.

Mm.

People get headaches like you have and have nothing. They just come, and they go away.

I see.

Where we sit right now is, there is a suggestion that you have a tumor of the pituitary gland.

Mm.

The most common one— ah, name, is called a chromophobe adenoma. You know, but the name is unimportant, except, as I say, it's not cancer.

Mm.

Now. If that's what you have, you're going to be operated on for it.

Mm.

Now, as scary as it may be, when somebody points at your head and talks about tumors, this is NOT a scary tumor.

I see.

I don't mean to say that on the way, you ain't going to get scared and worry and be frightened and have things go on, doctors and all that, because you are. Until this is all settled, you're going to get frightened, and nothing I'm going to tell you is going to make you less frightened, except, I'm telling you the truth. You're going to come out all right, if that's what it is.

Mm-hm.

All right? And I don't mean come out all right, I mean alive, I mean all right.

Mm-hm.

You know? Able to do your exercises, work, have sex, cook food, get mad at your children, and grow to be one hundred and ten years old like your mother. Nasty. With a nasty disposition.

(Laughs) OK.

So, but I— but we all— we both live in this world, and we know there's nothing scarier than somebody telling you that.

Yeah, I'm kind of— I'm kind of—

All right.

There's what you should be, and the— and the brain.

There are many worse things. Many worse things.

Yeah.

Now. My solution to scary things is go find out, you find out fast, and get it done, and get it over with.

That's what I want. That's what I want to do.

All right. OK. Now, let's see, wha— what questions do you have?

Questions? I ha— I don't know, right now. I cannot think about anything but that this kind of operation, after it's done, doesn't going to do anything about my ambition. I mean it's something very simple that, ah, I mean, they come out and that's it. It's not a consolance, though, I don't think.

It's scary. Never. It's scary.

It— It seems that anything that could go to your skull, you know, it's kind of a little hard, but, ah, I try not to scare myself. (Laughs)

Doesn't work so easy does it?

Yeah. I mean, this have to be removed, so, we, you know, fine! I believe in the doctors, you know, sometimes. You know, they do a lot of jobs; they— they open you, and they take the heart out of you. You know. And they remove it, and they put it back so I don't think this is going to be any harder than that.

Now let's make a pact between the two of us, please. And so that, you know, if you have a question, you ask me. If you want an answer, I'll give you an answer. Equally, if you don't want an answer, I won't give you an answer.

(Laughs)

OK. So finish your coffee, take a taxi right back down to the Health Service. That doctor's going to call me with his visual fields.

Mm-hm.

You'll call me maybe two o'clock.

OK. So I'm going to go do that.

And then I'll make your next arrangement— arrangements.

So are they going to, ah, do it today? They told you they're going to do it today?

Yeah. Because as I said, if they won't do it, I'll have it done somewhere else.

Ah, OK.

I'm not sitting around— I'm not sitting around that— so you can go through a weekend wondering, wondering, wondering? I'm not going through that—

OK, doctor.

Compare this conversation with our next example:

If she's supposed to have Parkinson's disease, how come she doesn't have any of the things that it says in the book she's supposed to have like tremors and—

Well, you can have Parkinson's without all of those things.

You can?

Sure.

What causes Parkinson's?

Nobody knows.

Nobody knows?

No.

Well, why— (laughs) should they, you know, you said—

If I started explaining to you, you won't understand, eh, a single word.

I believe it is unacceptable to tell a patient that something is "too complicated" to explain. If one really understands something, no matter how technical, it can almost always be stated in simpler terms.

If it is too complicated to explain, make a drawing. Patients do not usually want to know all the details of pathophysiology; they want to know the answers to their questions. To supply these answers, the physician must find out what the patient wants to know. Often an explanation can be given even to someone who speaks a different language. On one occasion I saw a Guatemalan woman with acute onset of bloody diarrhea. I wanted to use the sigmoidoscope and explained its necessity with a picture and some pointing. The typical lesions of amoebic dysentary were seen, and I drew some pictures to show what I had seen, and then a picture of the amoeba. Coming from a country where amoebic dysentary is a common disease, the patient understood. I was rather pleased that my less-than-artistic pictures did the job.

Sources of Information

If events are not explained and interpreted by the physician, the patient will provide meanings from other sources to fill the gap. Physicians must realize that they are only one source of information among many. The environment provides information. Patients in the waiting room of the radiotherapy department of Memorial Hospital have only to look around to get an idea about the gravity of their problem. Even if the patient is referred for a benign lesion, it would take considerable reassurance to overcome the effect of that particular context. The environment also includes newspapers, television, and magazines, all of which are filled with information about medicine and health. Information also comes from past experience and the experiences of family members and associates. Knowledge stored in memory, as well as beliefs about how the world works, also serve to give meaning to events. The person's needs, wishes, desires, fears, and fantasies—conscious or unconscious—are a source of meaning.

People also obtain information from their bodies. They have an expectation of how their body should look, feel, act, even smell. If the body changes, this is a potent source of information. It is a common error to belittle the patient's knowledge of his or her own body. If the patient thinks something is wrong, then something *is* wrong. The burden is on the doctor to demonstrate that the patient's perception of malfunction is incorrect (has no dire meaning), is accounted for by the physician's diagnosis or actions, or is in some other manner being taken into account. Again, if such information is dismissed as unimportant, some explanation is required.

Other people often tell a patient what to believe. One of the wonders of medical practice is the frequency with which people will believe their friends before they believe their physicians. In this next example we are privileged to hear such an information source: someone who had accompanied the patient to the office and was in the consultation room when the doctor prescribed prednisone.

So there should be no confusion unless I specifically tell you, "Stop the prednisone."

(Patient) Mm-hm.

prednisone is always tapered off.

(Friend) Her face will get rou— your face will get round, but don't get excited. Doesn't it.

What?

(Friend) From the cortisone.

No.

(Friend) No it will not?

You'll have no side effects at this—

(Patient) Ah, but it does, it—

No. Not like that, no.

(Friend) Just a little.

No!

(Friend) Not at all—?

Not at all.

(Friend) Very good.

More commonly, a physician sees a patient with severe poison ivy and decides that a short course of steroids is in order. The prescription is written, the method of taking it described in detail, and the person goes home. The friends or family then start telling the patient about ulcers, diabetes, cushinoid features, and osteoporosis, after which of course the patient will not touch the steroids with a ten-foot pole. I suspect that people give more weight to information about drugs and procedures provided by their friends and associates because they believe these people understand them, while the doctor's reasons for doing things (they may suspect) may be related to something other than only the welfare of the patient. In any case physicians are by no means the only source of information.

These other sources may provide misinformation, leading to wrong action, unnecessary fears, lack of compliance, and increased uncertainty. Correcting these errors in understanding is part of the doctor's job. Making fun of, or acting superior to, these meanings, merely belittles the patient; it will not correct misconceptions. We must start where the patient is and gently move back to the information the doctor believes to be correct. I am aware that patients and their families can be challenging, even infuriating, when they quote many different authorities in opposition to what you have told them. As difficult as it may be to remember, the important thing is not who is correct but how to get the patient to act in his or her own best interests. This requires avoiding disputes, determining what is best (and why), and then integrating these decisions with the patient's expressed needs and desires.

Vague Reference

"It looks as though we'll probably get out of the woods. Most of the bad stuff seems pretty well gone, and I doubt if we'll have trouble again for a while, if things hold the way they are." When physicians talk to patients with life-threatening disease, they commonly employ utterances like this. When asked, doctors expressed the belief that such sentences are reassuring. They are *not* reassuring, because they contain too much uncertainty. Faced with a potentially bad situation, most people fear the worst rather than expect the best. Therefore, when offered the kind of reassurance contained in such a statement, patients become more, rather than less, concerned. Here is what a patient might say, inwardly, on hearing the utterance. "It looks as though [*the doctor is not sure*] we'll probably [*but maybe not*]get out of the woods [*what does that mean?*]. Most [*but not all*] of the bad stuff [*what is "bad stuff"?*] seems [*is it, or is it not?*] pretty well [*but not all*] gone, and I doubt [*but the doctor is not sure*] if we'll have trouble again for a while [*how long is "a while"?*], if things hold the way they are [*what if they do not?*].

Patients may also receive, from such utterances, the idea that the doctor is afraid to tell them the bad news straight-out. Most speakers do not equivocate when they are certain of what they are saying, or when they have good news to relate. This next example is similar.

OK?
Tell me. Nobody tells me.
Well, what's do you want to know?
Wh- wh- wha- what that shadow is on th-the—
What shadow?
On-on the X rays.

There's no real change in your X rays.

Bu— j— there's still a-a-a tumor.

Well, some of it's shrunk down a little bit but, ah, you get scarring too, so, ah, you know, you've gotten a lot of medicine that we've give you for treatment.

But is, a-ah, th-the, ah,

I think we've made some progress. I'd like to give you some more treatment if I could. But what we've got to do is, we've got to wait 'til this fever is all squared away. OK?

I still have a little bit, huh?

A little bit. Not a whole lot, just a little bit. OK.

This physician was startled to learn that he had frightened the patient. He felt he had been absolutely straight and open. But rather than straight and open, he was vague. Every phrase contained the possibility of an opposite interpretation. The tumor has shrunk, but there's no real change in the X rays. Progress has been made, but more treatment is called for. It is essential to give information in a concrete manner. If the mass was smaller on the X ray, the doctor should have said so. You might object that a decrease in size on an X ray does not necessarily mean either smaller in reality or better prognostically. But these are separate issues. "The lump is smaller on the X ray." "Does that mean that I am better?" "No, it merely means the X ray shows improvement. However, smaller is better than larger, and after the medicine you have gotten, we expect the tumor to get smaller, and that is what the X ray shows." "Then, I could be *worse*." "Yes, or you could be *better*, which do you want to choose?" "I choose better." It is unusual for a patient to continue to pursue the worst possibility, but it does happen. Each question can be answered in a concrete and truthful manner, while still attempting reassurance. Each of the physician's hypothetical statements addresses a limited question and avoids large abstractions. Since we cannot be positive of the accuracy of our statements, you may wonder how we can reconcile the probabilistic nature of information, and decisions based on probabilities, with the need to be concrete when providing information. Let me repeat that *information is meant to meet the needs of the patient, not the physician.* A young woman was scheduled for surgery because of a mass in her pelvis. She complained to her father that nobody had told her anything. I was surprised, because I thought I had been very explicit with her. What I did not know was that her mother had taken diethylstilbesterol. Her fears about the possibility of cancer of the vagina, due to her mother's diethylstilbesterol exposure, were easily relieved. In discussions before surgery I had neglected to probe for her specific concerns. I could not know with certainty what the mass in her pelvis represented, but I could answer a specific question.

In this next instance the patient had been told about his Hodgkins disease while in the hospital. The conversation presented here took place after discharge from the hospital.

You just got out of the hospital, you've dealt with a— uh, unpleasant experience. Well, it's mostly called an unpleasant experience,

Yeah, well it's—

It's frightening. Getting your nose rubbed in your own mortality or the possibility of dying or being sick is no fun. You're entit— and besides which you handled the hospitalization extremely well, I mean, considering how scared you were before you went in.

Yeah.

You're entitled to— to— ah, still be focused looking at your belly button all the time.

(Laugh) Right.

But. What you're trying to do is get yourself, like I told you in the hospital, so you don't swing from blood count to blood count.

That's what I'm talking about.

You start doing that, I'm going to really give you a kick.

That's what I'm talking about. That's exactly what I want to get away from, so we—

It's hard to get away from it.

Right.

But, ah, I expect you're going to do very well. I can't promise you that.

Yeah.

That uncertainty—

Yeah.

—is the problem you must learn to live with. Uncertainty is your problem. You know you have Hodgkin's disease.

Yeah.

You know you can die from it in some w-way. I mean, all these things you know. You know you can also live—you know all those things, but you just don't know what's going to be.

Right.

So the problem isn't Hodgkin's disease. The problem's uncertainty.

OK. May—

Now, uncertainty is a thing. And it's THE THING you deal with, because no certainty can be introduced. In fact you can pick an object, like this paperweight, to stand for uncertainty. When you begin to get the "what if's," look at the object to remind yourself that uncertainty is a thing, just like the paperweight, and you can deal with it now because you have done it so well in the past.

Right, right.

All right? On the other hand, you can say to yourself, "How do I feel?"

Right.

"I feel fine." And then next you can say, if you get symptoms and they begin to do a number on you, you pick up the telephone and call.

Yeah.

I mean—

So, the way that I— so, that's the thing that I've probably said, and what you've said more clearly, is that's what the uncertainty is my enemy or obstacle or the thing—

Yeah.

—I'm dealing with.

Fear and uncertainty are the enemies.

Right.

They're a much worse enemy than Hodgkin's disease. 'Cause Hodgkin's disease is what it is and nothing more. It is what it is and nothing less, but it is—

Yeah.

—*what it is and nothing more.*

Right. Right.

But, fear and uncertainty are—vague, and—

Yeah, um, and, manifest themselves all sorts of tricky ways—

Yeah. It's also very usual for people to drive everybody around them crazy—

Right.

—*with their, "But— but, what's— do you think that's—what? What's that?"*

(Laughs) I'm trying to be g—

"That's a flea."

And then if they don't respond—you know, they have— they have to take some of it, the weight of it. You don't want to take all the weight of it, but given, um, that that's the thing that I'm fighting, er, that I'm contending with, ah, if I had just a b— I don't know what I'm trying to ask you. If I just basically knew the spectrum that I can swing in and not have to think about calling you, you know what I mean? I-I-I-I h— Does that make sense?

I can't tell you every symptom that's possible—

Sure. Sure.

—*or not. I just can't. My own guess is that you should have very few symptoms except some heart pounding, or nausea, or*

Mm-hm.

—loss of appetite—
Mm-hm.
—or less than optimal feeling of well-being.
Yeah.
And then you're off your course by Friday, so you're— you ought to feel pretty well until your next one starts, which is four weeks later.
Right. And, um, when I come and see you at the beginning of each cycle, you do certain procedures to check on how things—
Mm-hm.
—are going, right?
Mm-hm.
And, um, ah, what am I—so as you're saying, what you— what I want to try to not do is be waiting for that next one, and to hear you say—
Mm-hm.
Ah, the way I sort of see it is that between our meetings, which is once, twice a month, unless something particularly peculiar happens, I have to go with how I feel and—
Yeah!
—and know that under the canopy of knowing that everything that you and Dr. C—— have said and seen and showed me has all been what you've wanted to see.
Yeah.
Is that right?
Yeah. Yeah. It's doing just what it's supposed to do.
So—
You're doing well. The lymph nodes have melted away. Your fever is gone. You tolerate your chemotherapy without tearing your hair.
Yeah. True. True.
You didn't get— you didn't get vi— violently ill. Nothing bad happened.
Right. Right.
Right?
So, ah, yeah. So— I-I-I— what's always in back, I can tell just through our— my meandering and as I speak to you is that I— something in me is saying, "Yeah, but tell me that it's just absolutely fine forever." You know? That's why—
No.
Or exactly no.
No.
Of course no.
No I won't because I can't, and you know I can't.
Right.

And the minute I start going, "There, there, Paul" (laughs) *—then uncertainty will really enter your life because you won't be able to trust what I say.*

Right. Right. Right. So I can see that, heh— the uncertainty plays— as we talk— can I extend waiting a b—? (Laughs)

You had body fears before any of this ever happened to you, Paul.

Yeah. In a sense, certain things are removed. Ah, a— the kind of fears that I had before—say, in the six-month period before or nine months or year I knew what was going on. Now I know that these people are watching this condition very closely, and it's almost in a sense as though certain things you can STOP worrying about.

This example deals directly with the problem of uncertainty. It is not possible to remove the uncertainty that hangs over patients being treated for malignant disease. At first, they do not know whether their cancer will respond to therapy. If a response occurs, they wonder how long it will last. Their fears *cannot* be dispelled by telling them not to worry, or that everything will be fine. They all know better than that. Such words merely close off further conversation with the physician. It is vital to free patients from bondage to blood counts, X rays, or other test results. At each visit, the patient waits to see whether relapse has occurred or remission remains. Will the platelet count stay up? Is the shadow smaller on the X ray? If things are good, there is a brief period of elation, followed by anxieties that mount steadily until the next visit. This is an intolerable existence. Since the facts cannot remove these fears, how are they to be handled? First, note that these patients do not differ from the rest of us—we are all uncertain whether we will live through tomorrow. We who are not sick, however, are able to keep out of mind (deny, if you wish) the danger in which we live, for three related reasons. The first is that nothing has happened to remind us forcibly of our mortality. The second is that we maintain our focus on the other aspects of our lives— relationships, jobs, pleasures. The third reason is that no one keeps reminding us of the potential threats to our continued existence. Although physicians have no control over the fact that illness has endangered a patient's life, they definitely can influence the other two forces that keep the patient in constant fear.

Of primary importance is how the doctor views the test or X-ray result. If he or she acts as though it is all-important, then so will the patient. Often the doctor emerges with a great big smile, as if to say that all is well. But if the test results suggest return of illness, this can be read all over the doctor's face, and he or she suddenly becomes vague and evasive. As difficult as it may be, we must train ourselves not to drape crepe over our voices when discussing bad news and not to be all smiles when telling good news. We must learn an attitude of neutrality. If the patient is still in remission, that's fine; if relapse is signaled, then action will be taken. If death is inevitable, other actions must be taken. Of course we want our patients to remain alive and well, but whatever happens, the patients must know beyond any possible doubt that we will continue to take care of them. If you think that learning to act in this fashion is both difficult and counter to "normal" behavior, you are correct. The physician's job is to take care of sick people, whether they are going to live or die. Everything else is secondary. Odd as it may seem, when relapse occurs, patients sometimes act as if they have failed their physicians and are afraid that their doctors will be disappointed by them and leave them. Perhaps it is not so odd when you realize that doctors do sometimes get a magical feeling of power and pride when patients do better than expected—a feeling that is undercut by patients' relapses.

If it is clear to patients that their physicians' primary goal is to return them to function, reinforce relationships, and encourage the other activities of normal life, then patients will orient themselves more in these directions. When patients are recovering, part of the physician's function is to give permission to live a normal life. Active encouragement is necessary, because patients are fearful that their actions may bring a return of illness. Specific discussions of sexual activity, work, and recreation may be required. The single best way to deal with the uncertainties of serious illness is by concentrating on a return to normal function. As illustrated in the previous example, the patient has to be actively led away from focusing on his or her own navel (inevitable during serious illness) and directed, instead, toward the surrounding world.

Explaining the Meaning of Illness

We have met the patient in this next example. She changed doctors because she was concerned about the seriousness of her illness and believed she was not getting any explanations.

I'd like to find out what's wrong with my lower right, ah, side of me. I've been, in September it was, ah, uh, I complained that I still had some pain there after an illness I had in June, and I was fluoroscoped and told that I was— the right side of my diaphragm was paralyzed. The word later was changed to "immobilized," and subsequently a radioactive liver scan was given me. But my liver turned out to be all right, um. And as far as I know, not having had another fluoroscope examination since September, the symptoms are still there. Ah, however, this lower right pain—which is tolerable—if somebody says to me, "You're going to have that the rest of your life." Fine. I just want somebody to say, "That's it," or "That's not it," or "It'll go further," or "Here's a pill," or "Here's what it was."

I find you to be a healthy woman at this time.

Excellent!

And I don't believe that your pain comes from your diaphragm, although indeed it may, and I may be wrong. But let us review the evidence. Right? Why should a paralyzed diaphragm give you pain?

Beats me. There was some kind of—explanation about making my—

What was the explanation?

About under stress requiring me to breathe heavy, it would cause a cramp that made me feel the pain.

Yes, but, you see, what you describe is— doesn't make sense. Now, aerobic activity is aerobic activity. And the diaphragm, it moves in and out to get you air. See, it doesn't matter whether you're jarring or you're not jarring, or anything of the kind. It moves harder, the more air you need. If you're dancing vigourously, you are going to need just as much oxygen or air to breathe as when you run.

Mm-hm.

If you're doing equivalent exercise.

Mm-hm.

Now, you've been exercising for a number of years now, and you know what exercise feels like.

Mm-hm.

At equivalent exercise, you should have the pain. Right?

Gotcha.

Next, and you don't in aerobic dancing, but you do in the jarring motion.

I can get a stitch in my side. Whatever that is.

That's different, isn't it?

Yes.

That's a different feeling, isn't it?

Yeah.

All righty. Ah, next. You have sex. Most people breathe rather heavily when they have sex. You don't have pain unless your partner is in a certain position leaning on you.

Mm-hm.

All right, well, you're diaphragm doesn't know anything about that, all it knows is whether it's moving air in or out. It doesn't even get to join the fun, it just moves air in or out like a slave. Right?

Mm.

And you don't have pain when you're panting, but you do have pain when your partner is on top of you.

Mm-hm.

All right. Next. You have some trouble when you're upside down breathing but it's not clear to us why that is yet. Now, the diaphragm, upside down or rightside up just works just as well.

OK.

It really does. It's stupid. It doesn't know upside down from rightside up.

Mm-hm.

Um, and, ah, when you take a deep breath the diaphragm expands this way, right? That would be against gravity, upside down, and when you breathe out, the (blows out) diaphragm pushes up. Well, but that's against gravity when you're standing up. But you're able to push against gravity when you're standing up.

That's right. I can do it standing on my head!

Next, your diaphragm moves to the crudest clinical test. That is, when I percuss down and stop and ask you to take a deep breath and then percuss again, that diaphragm's moving. Um, now I can't be sure of that, and I'm going to go after that— that—

All right. And I'd rather hear that than, "We have a tin can fitted for you, to step into—"

Right, but, we're coming back to why all this has such scary meaning for you, but that's another matter. OK. So the next thing is that the clinical finding does not go with "paralyzed diaphragm." Your right chest moves just as well as your left chest. For a chest to move—well, you've got to move air. To move air in somebody, your a— the diaphragm moves up and down. Not always—the diaphragm can be— you know, can be troubled, not completely paralyzed. If the diaphragm is completely paralyzed, it's way up; it rests way up.

I see.

When a diaphragm is at rest, its— then the muscle is weak. It's all the way up at the top. Yours is not. Next, if it's your paralyzed diaphragm, why does it hurt back here? Hm?

Beats me!

When you get a stitch in—

With me—

—which we do think has to do with the diaphragm, although this is the weakest chain in the evidence—it hurts up here, not back there. Next, if it's your diaphragm, why does it hurt when I press on the left side of your— excuse me, on the right side of your back but not on the left? And why does it hurt along the same rib margins, which when followed down are where your pain is. What a coincidence that is. And not only that, it's not just the pai— the pain, it's reproducible. Every time I hit that same knot, it hurts. Right?

Mm-hm.

Now, one thing about your body is that it's lovely, and it is not filled with knots.

Mm-hm.

I mean, I have patients who you can't find a spot on them that doesn't— they're all uptight. But you're not. Your muscles aren't tight, it's not your way. Your head may be, I don't know about that, but you must—

Well, it comes and goes.

Uh, so, once again, that doesn't seem reasonable. Next, what illness paralyzes the diaphragm? The illness that paralyzes the diaphragm has to affect—there are illnesses that do it—but remember the diaphragm is paralyzed not because the muscle is affected but because the nerve is affected.

OK.

The phrenic nerve. Um, there sure are things to do it, but you haven't described one yet. Next, this illness that you had, with high fever, and so and so; a number of viral and other diseases produce inflammation of the liver. I think— somewhere in my hea— Dengue fever fits the description that you had, I think there is Dengue fever there, but I don't know—it doesn't much matter, you're over it now. Um, when I put it all together, I just don't have— come to the same conclusion. Now, ah, you may like my conclusion better, but the question is, why should you? The pain is there just the same. What's the difference?

The difference is that I know it's from a knot and it's not from being half-paralyzed and feeling "invalided" in some fashion.

Yeah, well wha— what— yeah, but pain is pain, so why is it any different, if it's your diaphragm or—

Well, because I can understand and deal with a knot and if it's a muscle knot. I assume that's what we're talk—

All right, supposing I said, "It's a muscle knot in the diaphragm." Well, your diaphragm is what it is and that's what it is. So what? Why can you deal with one but not with the other?

I could deal with it in my diaphragm too.

Yeah?

If you tell me there's a knot in my diaphragm, I know it's a muscle. I understand that! I don't—

Mm-hm, but why have you had trouble dealing with it up til now?

It's never been that until now. Up until now it has been, "Your diaphragm is immobilized. It is paralyzed." And—

What does that mean?

That's what I didn't know. If it's a muscle knot, I don't think you'd die from it.

I see. And a paralyzed diaphragm?

Who knows? If half of it's paralyzed, when's the other half going to fall?

Right. I see.

Do I ever— shall I ever go to Mississipi again? Was it my— the pool water? Was it the swamp? I don't— It's a very iffy—

All right.

—thing for a person who's been healthy—

Let me tell you something else.

Mm-hm.

Let us suppose for the sake of argument that indeed your diaphragm was "paralyzed" or "immobilized," which somehow seems to have a different meaning to you. Does it?

It does!

Why?

If my hand is paralyzed, I can't do this.

Mm-hm.

If it's immobilized because this is resting on it—

Mm-hm.

—as soon as this releases, I can do this again.

Mm-hm. So immobilized is better than paralyzed.

No. They just have different meanings to me.

All right. All right. Well, I do try to make the point to people who I teach that the words have meanings.

I think they do, and that's the only reason I repeated both words because two different doctors used two different words.

Of course— you— which side of creative advertising are you in?

Writer.

Yes. (Laugh) Words have special meanings, don't they? Hm? Ah, in any case, let us assume for the sake of argument, but I am not willing to concede that it is indeed the case for you. All right, but let us assume that the diaphragm was paralyzed because your phrenic nerve was in some way injured or even that the muscle was injured which gets remote past belief at this point. Ah, so what. Diaphragms only recover or stay the same. They don't get worse.

I didn't know that either. I wanted—

Did you ask the guy?

—that information. Yes.

And?

There was no answer. The answer was, "If you'll come back in December, we'll run through these fl— fluoroscopes and other things again, and it always seemed to stop once he was satisfied that my liver was OK.

Yeah. Well, you see, diaphragms aren't— is not a mysterious organ. I mean, it only can do certain—

No thrills.

—kinds of things. The body is the stupidest thing in the world. It can only do what it can do. Eyes can only see, they can't hear; and diaphragms move up and down their muscles, the phrenic nerve does its thing—everything does its thing. It

can't do just any old thing. It isn't that there aren't a lot of wonderful, wondrous things in the body. There are, but within limits. Right? Er, just within limits. And, ah, so even if that were the case, so it's the case. So big deal.

Fine!

But I don't think it is anyway.

That's fine with me.

However, that's all wonderful, but you're going get a block away from here and you're going to say, "But what about the pain?" Ahh— I tell you what we're going to do. I'll make a deal with you.

It's not pain I can't stand. I told you that earlier—

That's right.

—unless you've got your thumb in my ribs.

Right. I'm glad you can stand it. Ah, um, I'll make a deal with you. You get me some records, and sign your release, some films or a report. If it says the diaphragm ain't moving right, I want to repeat that and, ah, get it done relatively quickly,

Mm-hm.

And, um, I'll make the pain go away. I may have to stick a needle in that, put some Novocaine in there, but I'll make the pain go away. All right?

OK. I need one more thing. The name of a pediatrician.

Ah—

For my eight-year-old.

A pediatrician, hm?

I'm not dissatisfied with the one I have. It's just that I can't understand him.

This woman's search for explanation and meaning led her to change physicians. Indeed, she left the physician essentially because he used the word "paralyzed" when referring to her diaphragm. The point of the example is not that one doctor said the correct thing and another did not; it is that patients (and everybody else) hear *every word*. And every word has meaning. Hearing the word "paralyzed," the woman pictured herself in an iron lung! Drinker respirators are rarely used anymore, but she did not know that. One might ask how any physician could know that she was going to become so upset by a word. The doctor could not have known, but the opportunity did exist to ask questions, check comprehension, and expand the explanation. When presented with examples like this, physicians and even students say that they lack the time for lengthy explanations. A few minutes more, however, and the patient might have stayed with her original physician. And, with practice, explanations take less and less time.

The physician not only explained to the patient what he thought was the matter but laid out his reasoning. Further he used as evidence facts about how the body works which she herself could confirm or deny. Generally, such lengthy explanation is not necessary. In this instance, however, a previous physician's interpretation had to be countered and a less ominous diagnosis offered. (It is always more difficult to remove a serious diagnosis after it has been made. It is one of the foibles of humankind that bad news is more convincing than good.) Uncertainty is this patient's problem: she cares more about *knowing* than about how bad things are.

Let us consider another example.

You're a healthy lady with, ah, recurrent urinary tract infections, and I'll tell you what we're going to do about that.

OK.

The reason— yuh. The reason that happens so commonly in women is that the first infection, ah, does not entirely clear—

Mm-hm.

—um, and consequently every time the setting in which the urinary tract infection occurs again, it sort of— well, the nidus is there. The, ah, source is there for another infection.

Mm-hm.

Um, different people have different settings, but s— sex is a common setting. Um, but when it happens more and more frequently, less and less is required to make it start, and I'll show why that is. (Draws a picture on a pad) That's the mouth of the vagina, and this is the opening to the bladder, just above that. This is the vagina and then the rectum's right back there, and this is the pubic bone, and the bladder is like that and then the uterus is behind there, and so forth.

Mm-hm.

Now, the reason trouble happens is that, um, actually the urethra is not just a straight tube, as it is often pictured as being. But it's a tube with a lot of little glands that open on to it and make mucus, just the way the rest of the body lubricates its parts. And they're a special kind of glands, they're called the lacunar glands because of the funny shape they have, and so forth. And they all open into the urethra. Now, when you get an infection, commonly, it generally is really not a bladder infection. Although bladder infections certainly happen. Commonly, the infection starts in these little glands, and then they get infected. And when they get infected, it changes the glands a little bit, and so infection just sort of sits in them. And it doesn't drain well because of the shape they have. Y'see, when glands drain very well, infection is cleared quickly. Most mucus glands in the body are goblet cell glands. And they're in this shape, and they drain very well. Anyway, now if you can see the urethra is stuck between the vagina and a bone

Mm-hm.

—so whenever the vagina is somewhat traumatized, ah, then the urethra is too. And some of this infected stuff is pushed back up into the urethra. And it inflames, and it goes back into the bladder—

Mm-hm.

—and you get all the symptoms that you have. Now, the bladder actually is very resistant to infection. But, and so is the urethra—but once it's gotten one or two or three, it changes—and then it sort of holds on to the infection. And the secret of making it better is really not eliminating infection or a— bugs, because even though you get rid of the bug, another bug comes—it's ho— making it— keeping it free of bugs so long that it'll— it's allowed to heal.

Mm-hm.

And once it heals back to the way it was originally, then you don't get infections in the same setting that you got them before.

Mm-hm.

So, we're going to treat you for— for three to six months—

Mm-hm.

—and, ah, the treatment is very mild and um, wonderfully approved nowadays because what it is is a drug called Mandelamine, which is a urinary antiseptic. It's not really an antibiotic; it's a urinary antiseptic. And it works best in an acid urine, and we make the urine acid by giving you vitamin C, you see. So now you're now—Right?

Good! I like vitamin C. (Laugh)

All right, now you're going to be au courant because you're going to be getting vitamin C and the whole world loves that, so—

Here the patient is given a detailed explanation of why her cystitis recurs and why it is so difficult to eliminate. Why bother with so much detail? The physician wants her to take medication for six months. Would you take pills for six months without knowing why? It might be argued that merely the absence of recurrences would ensure compliance. Perhaps that would be the case for a brief time, but taking medication is bothersome, side effects are common, and the drugs are often expensive.

Another reason for so much detail is to disconnect the disease from the meanings that the patient may have ascribed to it. Illnesses like cystitis are often bound up with sexual activity and thus acquire meanings far beyond the diseases themselves. Patients frequently "take the blame" for the diseases, as though they were causing them intentionally: to avoid having sex, or to punish their partners, or to derogate their sexuality. On the other hand, the patient may believe the cystitis is a bodily expression of guilty feelings or other emotional states. All too often physicians agree with such interpretations, only on the basis of intuitive evidence. Although it seems unquestionable that such meanings may underlie or become associated with sexually related symptoms, rarely do such interpretations offer a substantial basis for treatment. The "hidden" nature of the female reproductive organ system (too many women still know almost nothing about their genitalia) and the lack of solid information about the nature of the disease makes it easier for patients to ascribe their illness to emotional motives. It must be remembered that our conception of anything includes some beliefs about where the phenomenon came from and what will happen as a result of it. This is as true of apples and unicorns as of cystitis. In the absence of factual knowledge other causes and results are applied which may be far from the truth. I usually attempt to dissociate the disease from its acquired emotional meanings. I do this for two reasons. The first is that if the patient is having emotional problems related to sex, I would like these problems to be made explicit and treated as such rather than bound up with an organic disease. The second reason is that I do not want to legitimate inadvertently the psychological interpretations that the patient is making, and thus seem to be confirming the fact that the patient feels guilty about sex when I do not know that to be true. (There are occasions when I believe that vaginal or bladder syptoms *do* come from psychological problems. In these instances I try to find out enough to confirm or deny the suspicion, so that a solidly based recommendation can be made for, or against, some straightforward psychological intervention.) By making the anatomical features of the urethra explicit, and by drawing pictures as part of the explanation, the whole problem is switched from hidden or unconscious processes to explicit understandings. As discussed in volume 1, the change from one dimension of meaning to another, from, for example, body sensation to cognition, can alter the patient's behavior in regard to the illness. The explanation becomes part of the treatment. It must be apparent that the effectiveness of these explanations does not rest on their being "true." Although I do my best to be correct in my anatomical and pathophysiologic descriptions (using textbook pictures and anatomical models where possible), error is inevitable.

In the next example, changing the patient's understanding of "addiction," and "painkiller," allowed her to obtain adequate pain relief.

There's three words you said to me and I keep them in my mind all the time: "Anticipate the pain. Anticipate the pain."
Right.
So I did that. I felt like I was going to get a pain here, and I— I quic— asked for the painkillers right away, you know? And gee, it was terrific. I didn't even have any pain at all.

Isn't that easier than—
It was terrific. (Laugh)

This patient was being taught how to handle her pain. An essential element was the need to change the meaning of addiction, since, as we saw earlier, she feared this more than she feared pain. Patients will remain in agony, hour after hour, rather than risk addiction—*even when they know they are going to die!* Reluctance to take medication may be appropriate in health, but in terminal illness exaggerated fears about drugs get in the way of achieving the most function possible. It requires considerable skill at juggling analgesics and the other agents used with them, such as the phenothiazines or hydroxyzine, to obtain maximum pain relief with minimum side effects. In order to teach those skills, doctors have to reeducate the patient by changing the meaning of drugs and their effects.

Just as fears about drugs may have to be overcome to achieve good symptom control, so too do patients sometimes have to be taught how to behave in the hospital, in order not to complicate the illness with fears linked to hospitalization. Physicians are so much at home in their hospital that they may not recognize what an alien and frightening world it can be. Sometimes even the hospital routine is a source of pain to patients, especially those who have difficulty coping with people in authority. Virtually everyone on a ward exercises some power over the patients. The authority is generally exercised for the good of the patient, but there are situations where a patient would do best to resist. In this next example the patient had to be taught to say "No." She was admitted to the hospital with carcinoma of the breast metastatic to bone and with long-standing chronic congestive heart failure from rheumatic heart disease. On the evening of admission she said that somebody had mentioned something about cancer and that she never wanted to hear that word again. She could not have been more explicit. During all of her care I never spoke directly about cancer, or about dying. Instead, we had conversations about her father's death, and what she had wanted for him, and how she felt about others. It is not necessary to be direct in order to accomplish the functions of information; metaphors and analogies will do the job, when necessary.

But, anyway (laugh)
All right—
Anyway, ah, I had— one of the doctors was in around eight o'clock

this morning, and I had it. And then they— they wanted to get me up. And I said, "Well, I have to have my medication first."

Mm-hm.

So they said I couldn't have it because it's every four hours on demand. So they made me wait an hour. And then, they didn't give it to me long enough. So I think that that was r—

What do you mean they didn't give it to you long enough?

Like, not long enough to take ef—

May I teach you some words?

Yes, sir.

The first and simplest words is, "No." "Ma'am, would you please get up?"

(Laugh) No.

No.

Oh.

"Oh but why not, Mary?" "I haven't had any pain medication, first. I've had a painful back longer than you have, and I know. And when I have my pain medication, and when I've had about forty-five minutes for it to act, I'll happily get out of bed, Nurse, and do whatever you want."

All right.

Please?

I didn't know that.

I'm telling you now.

All right, because I was going to ask you to change the hour and make it three hours on demand.

We'll make it three hours on demand. But.

But I'm to say "No."

Correct.

OK. Thank you.

The words, "No," or "No, thank you," or "Please not now," I th— um, any major, important thing that anybody wants you to do, they must have a good reason for it. And if it's a painful thing, they must have a good reason why it is painful rather than not painful.

Because I don't see that it does me that much good.

Lady, there is no need to have pain.

Mm-hm. Well, I know you impress that on me.

Well, but it's more than that. You see, people who have never had a lot of pain in their life, are not aware that pain is made up of both pain and fear of pain. And that people who have recurrent pain for a long time suffer as much from the fear of—as from the pain itself. And since all it takes is, "Oh, sure, Mary, we'll give your medicine now and come back in about an hour." That's all it takes.

Mm.

Right.

Mm-hm.

And then finally—

All righty, th—

—I want that—again and again and again. In the hospital or out, I want you to understand that you have more control over this situation and your body than you have any idea.

All—

Teaching the patient to say "No," strengthened the patient's relationship with the doctor—they became more of a team. This was accomplished by having the doctor ally himself with the patient against the other staff. Alliances of patient and doctor against the other staff members should be made with forethought and care. Often it is best to limit the conditions and be supportive of the staff in other regards. Manipulative patients can easily achieve the feeling of being in control by controlling the staff, playing one against the other. When this seems to be occurring, staff members would do well to discuss it among themselves and then decide on a common approach to the patient. It is important, however, to avoid punishing the patient.

Changing the World by Changing the Meaning

The next example introduces the final and, in some respects, most important point of the chapter. Patients may acquire many diseases and conditions for which we have no effective treatment. Indeed, significant numbers of our patients now suffer from degenerative diseases for which cure is not possible. Sometimes, however, when the disease cannot be treated effectively, therapeutic goals can be accomplished by changing the meaning of the disease. Not too many years ago, because of the lack of adequate therapy, the diagnosis of Hodgkins disease was a statement of impending doom. Today, on the other hand, we expect to cure the majority of patients with limited disease, and we even have much to offer those in advanced stages. The mood of a physician facing a patient with newly diagnosed stage IA Hodgkins disease reflects this change in outcome. The physician is optimistic, and the optimism spreads to the patient. The optimism does not arise because the doctor knows that *this* patient will be cured—that cannot be known—but rather because the *meaning* of Hodgkins disease has changed from an almost invariably fatal malady to one in which cure is possible. If the patient is discouraged, the doctor will hasten to explain how much better things have become nowadays, thereby changing the meaning for the patient. The change in the meaning of Hodgkins disease, which can now be shared by both doctors and patients, is part of the invisible context of treatment, and this context influences the course and outcome of therapy. One does not have to look to arcane explanations of the effect of the mind on the body for the influence of meaning. How actively the patient will cooperate—show up for treatments, comply with medication regimens, report symptoms accurately—is determined, in part, by how hopeful the patient feels. It is also true that a doctor pervaded by a sense of hopelessness may avoid the patient, not try as hard, or even subtly discourage the patient from further therapy.

Cancer offers extreme examples, in which a change in meaning can profoundly influence a patient's well-being. In the next example the patient had just returned to her room after an exploratory thoracotomy. Resection of her carcinoma of the lung was impossible because the tumor was bound to the chest wall and to the hilum. Ordinarily I avoid lengthy explanations so soon after surgery because the patient's comprehension may be impaired by medication. In this instance it was necessary because she pursued the topic.

What else. And I've been trying to do everything—that I'm supposed to do.
So far it looks to me like you are, too.
OK. Well, my family thought I looked fantastic. They said—
You do!
They said, "You weren't in the hospital for h— surgery!
Yeah, Dr. O—— said, "Ah! Wait'll you see her! She looks like before the operation." You look good.
May I ask you a question?
Sure can!
Was it a very, uh, uh, extensive surgery?
No.
No?
No. It wasn't. It was as extensive as it could be.
—What does that mean?
Well, you got a tumor there that's gonna to be, ah, treated with radiation. And it's not going to be, removed. So it'll stay right where—
Another tumor?

Not "another tumor." What does "another tumor" mean?

I thought he cut it out?

No, uh, in this position it gets treated with radiation.

Benign?

Hm? No—it's not. But it is going to get a good response to radiation. You're going to get a good response. You're going to be well again. You're going to get a good response. It doesn't matter that it hadn't— couldn't be cut out. It does not matter. See, one thing about it is— the one thing about— about when I'm truthful with people, you have to listen to all of it because when I say things that don't fit what you believe or know, you got to know I must have a reason for saying it. Now, uh, I'll go over the details with you but— when— after you get out of the hospital. You're going to have to get X-ray treatment to that area, and that's going to scar that tumor down like a walnut. But because of where it is, operating on it would do you more damage than taking it out would do you good.

I thought it was all the way up here.

It is. It is all the way up there, but it's attached against th— the top of the chest wall. To d— to take it out would require taking out a part of the chest wall, and that would be really a lot of trouble. And it does just as well treated with radiation.

It does?

Yes Ma'am. Yes Ma'am. It's that kind of a tumor. Radiation will just punch it down.

But it is a malignancy.

Mm-hm. Listen, Ma'am, you're going to get treated. You're going to be well. You're going to return to work. That's what's going to happen to you. You're going to go right back out and be alive and kickin' again.

—OK!

Because the thing that'll keep— I mean, you know, radiation is what, eh—it's an annoyance to keep having to go to a doctor's office to get radiated—it takes over a four week period. But, my last patient with the same thing, worked while she was gettin' radiated.

Disappeared?

Hm? Oh, yeah. Shrinks right down.

And they don't come back?

I can't promise you that.

Ohhh. (Sigh)

But listen to me, Bette, listen to me. You're going to come out, and you're going to be alive again. You're not a dying lady, and you're not about to die.

I'm not?

No. I'm telling you the God's honest truth. You are not a dying lady, and you're not about to die. You're going to get treated, and you're going to go back to work. You're going to be a living person. I mean, and, and, you know, those words— I'm able to say those words. And if you were a dying lady, I would have to deal with that with you and take care of you, would't I? I mean so, when I tell you, this, it means that's m— my problem with you, with getting you—with getting you to do that. In other words, the tumor, it's gonna do what it's gonna do, because the radiation's gonna blast the shit out of it—

But it didn't move or anything in all this time . . .?

No. No. No, it's that kind of a tumor. You see, and I'll draw you a dia— I won't take the time now, but I'll draw you a picture tomorrow when you're feeling better, and so forth, and you know, and show you where and why and what it does and why it, you know, why we leave it behind and why taking it out would be more trouble to you than— than leaving it in and what the radiation does—

So nothing was removed, huh?

No. Biopsy. That's all. To make sure.

I'm a little shocked. (Laughs)

Yes I know you are. Should I have told you a lie?

No.

Hm? When you got the radiation, and I told you a lie, what would you think you were getting the radiation for? A cold? Only you'd know that if I— if nobody's telling you, it must be terrible. And it isn't. You're going to get treated, you're going to get well again. No, I don't promise you won't get a recurrence, I can't do that. But I can promise you're going to be well and working again. You're going to be— you are not a dying lady. And my problem will be not the tumor.

Hmm.

My problem will be you! You know, you can be a dying lady with a living body. I ain't having that. You know what I mean? You can be tied to your tumor when your tumor is already gone and you're already—you're still tied to it. I'm not having that, Bette. I'm about living and health and doing and families and gettin' on about your business, because there's lots of things you don't know about medicine—I'm going to teach you about 'em, because they— they'll help you in this instance, 'cause you— all I have in you is preconceptions and misconceptions and "uncle this one and Sammy that one and Bill so-and-so." I can't work with that. I have to know about it, so I can answer questions. But we're concerned not about all that and this one and your husband's ex-wife and th— whole prev—you know. I'm not concerned about them, except as they, you know, except as they feed your fears. 'Cause they do, that's how we get our fears. I'm concerned about yours and what it's going to do and how it's going to do. And so you know about that so that you're freed of it. Otherwise you're tied to it? Understand me?

Mm-hm, yes.

That conversation took place in April 1975. In August 1975 the patient developed weakness of her left arm and leg and was found to have a large solitary metastasis to the brain which was successfully operated. She returned to work and to her life in all its dimensions. She remained well until fall 1980, when she died from a recurrence of tumor. She lived with a sense of control and was not a "dying woman" during those five years.

In order for that to be possible, it was necessary to counter all the meanings she already attributed to cancer and to radiation. How does one know what these meanings are? These meanings are, above all, the myths of our culture and can be learned by questioning the patient. The task is not merely to tell the patient that she is not a dying person; you must convince her that she can return to her previous life, *and then show her how.*

All around the patient are messages to the contrary. For example, consider what occurs when this patient goes to the operating room and is found to have inoperable cancer. Even before the patient returns to her room, the messages begin to change. A nurse says, "We heard the terrible news about Mrs.———. She's not going to live, is she?" People begin to enter her room with their faces slightly averted and start to address her in funereal tones. Without a single explicit word the meaning of her disease will become clear to this woman. She knows she is doomed. Nothing has to be openly acknowledged: our lives are lived in social interactions, and we respond to them as synchronously as a dancer in the corps de ballet. The patient assumes her part in the ballet: the role of sick and dying wife and mother. We do not understand all the subtleties of the performance because sufficient attention has not been paid to this ceremony. We know, however, that the scene, and the behavior of the performers, would be different if it were a young businessman or an old person. The details are not important here; what is important is that the scene will change if the patient refuses to play her part. If she can be convinced not to play the dying woman in her own eyes, she will not be considered dying in the eyes of most others.

The use of information in medicine to change the meaning of events, feelings, and diseases is by no means confined to cancer. Symptoms and disease states associated with aging offer additional examples. A patient in his sixties developed pain and swelling in his knee which was shown to be due to osteoarthritis. Although the symptoms subsided rapidly, the patient began to cut back on the long walks he had always taken and to become depressed. In discussing the problem with the patient, two things became clear. First, he considered the difficulty with his knee to be a sign of aging; it heralded his now inevitable, and soon to be progressive, decline. Second, he believed that the knee had only a certain amount of function left within it, so that if he used it less (walked less far), it would last longer. Both are common beliefs. In fact, however, although osteoarthritis is unquestionably often related to aging, neither idea is correct. On examination the troubled leg showed decreased muscle mass, which was corrected with exercise, physical therapy, and an orthotic in his shoe. The disability of the knee was then used to demonstrate to the patient that he must consciously care for his body and that this required more exercise rather than less. In the same way he was disabused of his belief in the ineluctable deterioration of aging, thus preventing the fear from becoming a self-fulfilling prophecy. It is true that when patients like this restrict their activity out of desire to conserve their remaining function, their functional reserve decreases. It is their self-imposed restriction, rather than the aging process alone, that is responsible for the deterioration.

Many aspects of the aging process, from mental function to sexuality, are open to similar changes in meaning. Reeducation can transform a patient's outlook and function and may even affect the underlying physiological processes. And so it is with virtually all of the chronic illnesses. Will the diabetic patient live in terror of gangrene and restrict activities for fear of injuring the feet or, with that concern reduced by adequate information, go on to a more active and healthy existence?

You may object that I am merely talking about health education, the value of which is obvious. I am indeed discussing health education. It is strange how ineffective health education has been, however, in changing health behaviors such as diet, exercise, or cigarette smoking. The reasons for the failure are not clear, but health education, as it is generally considered, differs profoundly from the reeducation I am now discussing. Each of the instances I have presented is related to a specific fear or concern of a patient that has been elicited at the time of some illness or symptom episode. The patient is motivated by the immediate concern to pay attention and also, perhaps, to change. The physician has a chance to bring about a more general change in belief (about aging, for example), by presenting new ideas in the context of a concrete and immediate problem.

There is another difference between the process I am discussing, and "health education." I am suggesting that, by changing the meaning of an illness, altering its significance and importance to the patient, the patient may get better, even though the illness itself cannot be treated effectively. The change in outcome is brought about, at least in part, by changes in the patient's behavior toward the illness, resulting from his or her new understandings. Earlier I noted that one does not have to look to esoteric views of mind-body interactions to see how a change in meaning has an effect on the patient. Recent studies in neuroendocrinology, however, provide a concrete basis for understanding the way a change in meaning may lead to a change in physiological function. The simplest example is fear. There seems little doubt that chronic fear (as might be experienced by someone who believes that life-threatening disease is present) has an effect on endocrine function. If this is the case, removing that fear, by changing the meaning of the disease, would also influence endocrine function. I chose the most obvious example to make the point, but whether someone is enchanted, terrified, saddened, or exalted by events is a function of the meaning assigned to those events. All of these emotions have physiologic correlates. Thus, from the level of psychosocial function to that of the hypothalamus, the outcome of an illness is influenced by its meanings to a patient. And these meanings can be influenced, expanded, altered, through the efforts of the physician.

We have traveled a long road in these two volumes—from the intricacies of the function of language in medicine to its use as a most sophisticated diagnosticate and therapeutic modality. I believe you will understand now why I stated in the introduction that the spoken language is the most important tool in medicine. The knowledge that can be learned from effective communication with sick persons, and the healthy also, is not only essential to their care but offers the opportunity of returning to medicine all the human dimensions of our patients (and ourselves) that seem so threatened by technological imperatives. Here lies the true partner to medical science and technology and the opportunity for a balanced medicine. I believe that the examples offered in these two volumes (as well as our everyday clinical encounters) make it abundantly clear that, in terms of diagnosis and treatment, a disease cannot be understood apart from the patient in whom it occurs. As your skills in the spoken language increase, you will find that they open to you an ever-increasing understanding of the world of sickness and sick persons—the exciting domain of clinical medicine.

There is more to it, however, than diagnostic accuracy and therapeutic effectiveness. As desirable as those goals are, there is another reward for the mastery of effective communication with patients—lifelong challenge and satisfaction in clinical medicine. There comes a time for almost every physician when the next case of, say, congestive heart failure is not particularly challenging or, sad to say, even interesting. But the next *patient* in heart failure—that is another matter. Every patient is challenging, interesting, and different, and every sick person offers not only the opportunity to help someone be better but the chance to know more about the endless mystery of sickness and the human condition.

Suggested Readings

Abercrombie, David. *Elements of General Phonetics*. Edinburgh: Edinburgh University Press, 1967.

Austin, J. L. *How To Do Things With Words*. New York: Oxford University Press, 1973.

Austin, J. L. *Philosophical Papers*. New York: Oxford University Press, 1970.

Bakan, David. *Pain, Disease and Sacrifice*. Chicago: University of Chicago Press, 1968.

Balint, M. *The Doctor, His Patient and the Illness*. 2nd Ed. London: Pitman Medical, 1964.

Bauman, R., and Sheizer, J., eds. *Explorations in the Ethnography of Speaking*. New York: Cambridge University Press, 1974.

Brown, Gillian. *Listening to Spoken English*. London: Longman, 1977.

Bruner, J. S. *Beyond the Information Given*. New York: W. W. Norton, 1973.

Byrne, P. S., and Long, B. E. L. *Doctors Talking to Patients*. London: Her Majesty's Stationary Office, 1976.

Cassell, Eric J. *The Healer's Art*. New York: Lippincott, 1976. Penguin, 1979.

Cassell, Eric J. *The Place of the Humanities in Medicine*. Hastings-on-Hudson: The Hastings Center, 1984.

Deutsch, F., and Murphy, W. F. *The Clinical Interview*. New York: International University Press, 1960.

De Villiers, J., and De Villiers, P. *Language Acquisition*. Cambridge, Mass.: Harvard University Press, 1978.

Enelow, A. J., and Swisher, Scott N. *Interviewing and Patient Care*. New York: Oxford University Press, 1972; 2nd ed., 1979.

Fodor, J. A., Bever, T. G., and Garrett, M. F. *The Psychology of Language*. New York: McGraw-Hill, 1974.

Gordon, George N. *The Languages of Communication*. New York: Hastings House, 1969.

Halliday, M. A. K., and Hasan, R. *Cohesion in English*. London: Longman, 1975.

Halliday, M. A. K. *Language as Social Semiotic*. London, Arnold (Publishers), 1978.

Hardy, W. G. *Language, Thought and Experience*. Baltimore: University Park Press, 1978.

Hare, R. M. *The Languages of Morals*. New York: Oxford University Press, 1969.

Hayden, D. E., Alworth, E. P., and Tate, G. *Classics in Linguistics*. New York: Philosophical Library, 1967.

Hymes, Dell. *Foundations in Sociolinguistics*. Philadelphia: University of Pennsylvania Press, 1974.

Howell, R. W., and Vetter, H. J. *Language and Behavior*. New York: Human Sciences Press, 1976.

Jokobovits, L. A., and Miron, M. S., eds. *Readings in the Psychology of Language*. Englewood Cliffs, N. J.: Prentice-Hall, 1967.

Judge, R. D., Zuidema, G. D., and Fitzgerald, F. T. *Clinical Diagnosis*. 4th Ed. Boston: Little Brown, 1982.

Labov, William. *Language in the Inner City*. Philadelphia: University of Pennsylvania Press, 1972.

Labov, William. *Sociolinguistic Patterns*. Philadelphia: University of Pennsylvania Press, 1972.

Labov, William, and Fanshel, David. *Therapeutic Discourse*. New York: Academic Press, 1977.

Ladefoged, Peter. *A Course in Phonetics*. New York: Harcourt Brace Jovanovich, 1975.

Lain-Entralgo, Pedro. *Doctor and Patient*. New York: McGraw-Hill (World University Library), 1969.

Lain-Entralgo, Pedro. *The Therapy of the Word in Classical Antiquity*. New Haven: Yale University Press, 1970.

Laver, John, and Hutcheson, Sandra, ed. *Communication in Face to Face Interaction*. Baltimore: Penguin, 1972.

Lyons, John, ed. *New Horizons in Linguistics*. Baltimore: Penguin, 1970.

Lyons, John. *Introduction to Theoretical Linguistics*. Cambridge: Cambridge University Press, 1968.

Lyons, John. *Semantics*. Vols. 1 and 2. Cambridge: Cambridge University Press, 1977.

Lyons, John. *Language Meaning and Context*. London: Fontana, 1981.

McIntosh, J. *Communication and Awareness on a Cancer Ward*. New York: Prodist, 1977.

Miller, George A. *Language and Communication*. New York: McGraw-Hill, 1951.

Miller, George A., and Johnson-Laird, Philip N. *Language and Perception*. Cambridge, Mass.: Belknap Press, 1976.

Ozer, Mark N. *Solving Learning and Behavior Problems of Children*. San Francisco: Jossey-Bass, 1980.

Percy, Walker. *The Message in the Bottle*. New York: Farrar, Straus and Giroux, 1978.

Reiser, David E., and Schroder, Andrea K. *Patient Interviewing*. Baltimore: Williams and Wilkens, 1980.

Richardson, S. A., Dohrenwend, B. S. and Klein, D. *Interviewing*. New York: Basic Books, 1965.

Rochester, Sherry, and Martin, J. R. *Crazy Talk: A Study of the Discourse of Schizophrenic Speakers*. New York: Plenum, 1979.

Schiffer, S. R. *Meaning*. Oxford: Clarendon, 1974.

Searle, J. R. *Speech Acts*. Cambridge: Cambridge University Press, 1969.

Watzlawick, Paul. *How Real is Real*. New York: Vintage Books, 1976.

Watzlawick, P., Beavin, J. H., and Jackson, D. D. *Pragmatics of Human Communication*. New York: W. W. Norton, 1967.

Index

Abortion, 100–101
Art of medicine, 1–3. *See also*
  Information.
  four aspects of, 2
Attitudes
  need for physician to maintain neutral
    attitude, 176
  of patients about causes of illness, 37;
    about pain, 6; about self, 6
  of physicians about patients, 87, 88

Behavior, self-destructive, 135–136
Beliefs. *See* Attitudes

Clinical medicine
  knowledge of human biology in, 108
  vs. medical science, 5
Communicating with patients
  in clinical setting, 10
  kinds of information revealed in, 4
  listening as an aspect of, 3
  need for explanation in, 167–168
  in practicing the art of medicine, 2
  style in, 157
  use of verbal and nonverbal language
    in, 3, 4
*Communication and Awareness on a Cancer
  Ward* (McIntosh), 148

Denial
  of alcoholism, 91
  as a symptom, 25
  time as a mechanism of, 28
Diagnosis. *See* Medical diagnosis
Disease, human. *See also* Illness
  asthma, 52–54, 56
  carcinoma of the lung, 188–190
  carcinoma of the pancreas, 80–81
  congestive heart failure, 32, 55
  cystitis, 182–183
  diverticulitis, 20
  gastroenteritis, 15, 17
  histiocytic lymphoma, 90–97
  Hodgkins disease, 172–175
  hypothyroidism, 70–74, 139–141
  infected sebaceous cyst, 19
  inflammatory disease of the rectum,
    45–51
  migraine headache, 16–17, 20–21
  patients' concerns about, 97
  pericarditis, 82–85
  thrombophlebitis and pulmonary
    embolis, 12–13, 17, 63–65, 109–110,
    131–133, 142–143, 162–164
  ulcerative colitis, 110–112
Doctor-patient relationship
  dyadic nature of, 4
  example of communication in,
    152–154
  use of toward therapeutic ends, 2,
    150

Experience. *See also* Meaning
  assignment of cause to, 29–30
  assignment of meaning via experience
    of others, 27
  patients' methods of reporting, 31–32
  patients' need for aid in relating,
    36–37
  space-time dimension of, 27–29

Fatal premise, 30

History. *See* Medical history; Medical
  history, past; Medical history,
  personal; Medical history, taking of